PLASTICS
DESIGNS AND MATERIALS

STUDIO VISTA LONDON

PLASTICS
DESIGNS AND MATERIALS

SYLVIA KATZ

To all my friends.

Acknowledgements
I would like to express my gratitude to all those who have provided assistance, information, contributions and personal encouragement, and in particular to: Richard Chambers; J.H. DuBois; Harry J. Earland; George Griffin at Brunel University; William Harvey; Vivian and Tim Head; Neil Hornick; Roger Newport; Robin Penfold; P.J. Povey, curator of the Post Office Telecommunications Museum; The Rubber and Plastics Research Association; John N. Ratcliffe, Secretary-General of the Plastics and Rubber Institute; Dr C.A. Redfarn; Martyn Rowlands; Mike Seabrook of John Dickinson and Co. Ltd; Fred Scott; Robert Silberman; John Walker; Albert H. Woodfull; and thanks to Henrietta Pasley-Tyler for secretarial perseverance.

The trade-names listed at the end of each section are not up to date, but are intended to be used for historical reference.

Many sources of information were conflicting, and the author would be pleased to receive any further information to help establish the facts.

A Studio Vista book published by
Cassell & Collier Macmillan Publishers Ltd,
35 Red Lion Square, London WC1R 4SG
and at Sydney, Auckland, Toronto, Johannesburg,
an affiliate of
Macmillan Publishing Co. Inc.,
New York

ISBN 0 289 70783 8

Designed by Sandra Schneider and Anthony Cohen
Set in 10/12pt Monophoto Plantin by Rembrandt Filmsetting Ltd Watford
Printed by Sackville Press Billericay Ltd.
Bound by Webb Son & Co, Ltd.

The urea-formaldehyde plaque presented to all winning designers and material suppliers in the first Modern Plastics Competition 1936. Moulded by the Chicago Molded Products Corp. in one of the first beryllium-copper moulds. A plastics trophy for a plastics competition. (*Modern Plastics*)

". . . I'm sure that love
Will never be
A product of
Plasticity."

Frank Zappa. *Plastic People*, Third Storey Music Co., 1966.

Mr McQuire: *I just want to say one word to you. Just one word.*
Ben: *Yes, Sir.*
Mr McQuire: *Are you listening?*
Ben: *Yes, I am.*
Mr McQuire: *Plastics.*
Ben: *Exactly how do you mean?*
Mr McQuire: *There's a great future in plastics. Think about it. Will you think about it?*
Ben: *Yes, I will.*
Mr McQuire: *Enough said. That's a deal.*

(*The Graduate*, director: Mike Nichols; script: Buck Henry; United Artists, USA, 1967)

'Plastics should be sold for what they are—perfected industrial raw materials.' (*Plastics and Moulded Products*, Neckwear Publishing Co., New York, 1933)

'That's not plastic, Madam, it's Bakelite.' (a market stall-holder, 1975)

Contents

Introduction

'The Age of Plastics'

Many people, especially journalists and commentators, seem to think that we live in 'The Plastic Age'. In fact, every decade since the modern plastics industry began in the 1920s has been hailed as 'The Plastic Age'. By right that title is far more appropriate to those early years; new materials were being discovered at an increasing speed and the foundations of the industry as we know it were established. In 1925 *Plastics* had been founded in America, the first plastics journal in the world (later named *Plastics and Moulded Products* and now *Modern Plastics*), followed by a stream of journals in Europe. In May 1930 the first symposium on plastics was held at the Grand Central Palace, New York, attended by 608 people. In London in the same year, one hundred men convened to form a plastics club with the aim of increasing an interest in plastics in England. Finding the word 'Club' somewhat unprofessional, they changed it to 'Institute'; a committee of seven was appointed, and the Plastics Institute was established, now the Plastics and Rubber Institute.

Those early days reverberated with the cries of patent-filers and litigants, a sure sign of a healthy competitive industry. However, as *Modern Plastics* commented, progress benefits better when patents are licensed so that know-how can be shared rather than their being contested in lengthy disputes.

In the 1940s the attitude to plastics was no less bewitched: 'There grew up a very general idea that we were on the verge of a "Wonder Age of Plastics",' wrote the Council of Industrial Design in their 1948 issue of *New Home*. This view sprang from the war years, when amazing feats were achieved by plastics. The war nurtured the idea of plastics as 'wonder materials' capable of solving all our problems. Three decades later French retailers are again announcing that *'le plastique est roi'*.

However, as Dr Chilver, Vice-Chancellor of Cranfield Institute of Technology, has said, if any one thing characterizes our age, it is communications, not plastics. The Stone and Bronze Ages are so called because most implements were made of those particular materials. The same cannot be said of plastics today, because plastics are but one of the whole range of necessary materials. Plastics are yet to come of age—an event that will happen, according to J. Harry DuBois, in 1983 when the volume of plastics will equal that of all metals. It is estimated that one third of all chemists and chemical engineers in the West are now engaged on the development of plastics. In 1975 the Science Research Council (SRC) announced the establishment of a Polymer Engineering Centre, to coordinate both processing and design in plastics, while at Lewes Prison, in Sussex, sewing mailbags has been replaced by plastics moulding. Hopefully, however, there will never be a 'Plastic Age' in the proper sense of the term.

The asset of plastics is that, unlike precious metals and stones, it is a universal material, available to everyone. Synthetic resin is transformed into articles that can be exclusive and expensive, or cheap rubbish. Plastics can also be superfluously decorative, but at the same time, according to Charles Spencer, can fill a social need: 'Perhaps there is something in plastics, as a kind of democratic material, capable of the most mundane and the most serious uses, identified with the "space" age, which promotes a down-to-earth, social conscience' (*Art and Artists* London, September 1967). But as Alan Glanvill pointed out at the Society of Industrial Artists conference in May 1975, plastics have suffered from the success of fulfilling this social need, for they have thus descended the social scale in the mind of the user.

This book sets out to explain the difference between the major types of plastics, their chemistry and how they are processed. It illustrates the history of plastics and traces their use and development, identifying their properties and limitations.

However, this book is not an exhaustive technical survey of modern plastics. There are thousands of such books available for the avid reader to study. But there are few books which set the history of plastics materials in a perspective comprehensible to the non-specialist, relating them to familiar trade-names. Journals, shops and even researched exhibitions in museums do not inform the public of the type of plastics that are on display. There are at least sixty different plastics in current use, and the lack of information must surely result from ignorance rather than from a wilful desire to confound.

In 1948 many large London stores were training their staff to answer queries from customers about the plastic objects they retailed. There has been no progress since then; things are little better than in 1956, when E.G. Couzens and V.E. Yarsley wrote that, as far as official statistics were concerned, plastics had the misfortune of being considered almost anonymous materials. There seemed to be an unnatural reticence to accord plastics an individual status in official records, they wrote. Even now, 80 per cent of plastics are used in unseen applications.

For too long plastics have been shrouded in a technology that has perpetuated an aura of mystery. 'Made of plastic' on a label is about as helpful as a tag on a gold or silver bracelet saying 'made of metal'.

Plastics—'Cheap and Nasty'?

'I was born with a plastic spoon in my mouth . . .
Substitute your lies for fact, I see right through your plastic mac.'
('Substitute' by Pete Townshend, 1966)

Derived from the Greek adjective *'plastikos'* meaning mouldable, the Plastic Arts have traditionally signified crafts which employ materials capable of being moulded, such as metal, clay, plaster and glass. 'Plastic' in this classical sense has meant three-dimensional, formed, plasmic; hence the phrase 'God the great plastic Artist (1741)' *(Shorter Oxford Dictionary)*. We also apply the adjective 'plastic' to materials which are very flexible in their finished state.

The word 'plastics' in the plural, can be used as both noun and adjective. The word was officially recognized in 1926 as a collective noun describing a group of materials capable of plastic deformation. Its adjectival use was accepted by the British Standards Institution in 1951. Thus, 'The Plastic Age' should strictly speaking be 'The Plastics Age', but as 'plastic' is now the commonly used form it has been adopted throughout this book. However, the meaning of the word 'plastic' with which we are all most familiar is to denote something fake and synthetic. The problem with plastics, as anyone concerned with the manufacture, design, or the retailing of plastic goods will endorse, is the myth that still exists that they are a cheap substitute for something better.

Because plastics have suffered from the success of widespread proliferation, which has cheapened and debased them, the mistakes that have resulted merely reinforce the poor public image. Even now shoppers still ask: 'Is this real or plastic?' as if plastics could never be real. Semiotics, the general science of signs, points out that the more often a certain word or phrase is used, the more its symbolic meaning is reinforced, until it is finally accepted as a cultural cliché and is absorbed as a myth. This has happened with the words 'cheap plastic', which have been repeated so often that 'cheap' and 'plastic' have become synonymous, although plastics are becoming real materials in their own right for which there is no substitute.

We can see what plastics mean to our present society by looking at how the underlying myth is expressed through the verbal and visual use of plastics. For example, the phrase 'plastic people' has now become current, stemming from when

Many labels have been attached to the 1930s, from the 'Electrical Age' to the 'Jazz Age'. Posters for the *L'Age des Plastiques* exhibition held in the Maison de la Chimie, Paris, 1939, overshadowed those of all the normally more popular Parisian sporting events. (*Plastics,* Temple Press)

Frank Zappa sang in 1966: 'We are confronted with a vast quantity of plastic people' ('Plastic People', Third Storey Music Co.). The illustration shows another example. 'Behind Monroe's haunting smile lies the tragedy of America in the plastic age,' describes a false smile for the camera, immortalized eternally on celluloid film. But it haunts us because behind the smile lies hidden the 'real' person she once was before being moulded (like plastic) into a superstar. By the use of the word 'tragedy', the plastic age is seen as an unfortunate occurrence which leads ultimately to destruction.

Antagonism towards plastics is so established and its potential effect on marketing so heavy that manufacturers of modern paints dare not print the word 'synthetic' on their tins although most paints now are synthetic. Instead they use euphemisms like 'Silthane Silk' or 'Vinyl Satin'.

The origins of the present-day connotations of the notion of cheapness can be seen in the history of plastics. Crisis points in history, when the world's supplies of natural materials have been threatened by wars, population explosions or economic change, have often been a spur to technological development. The origin of our modern plastics industry stems from such a crisis in the 1840s when the population increase strained the supplies of natural materials such as silk, rubber and ivory. Those materials were already becoming increasingly expensive. Celluloid, the first man-made plastic, was originally patented in America in 1869, as a cheaper substitute for ivory billiard balls. Casein, which was patented in 1897 in Germany under the trade name 'Galalith', developed originally as a cheaper, substitute material for horn and silk. Phenolic resin was often referred to as 'imitation amber'.

World War I created renewed demands on all natural resources, over which governments assumed control, and speeded up the development of synthetics.

Before 1914 Germany had pioneered synthetic rubber, and by the outbreak of World War II had prosperous PVC and polystyrene industries. World War II stimulated research in all fields of polymer chemistry. PVC, polythene and 'Perspex' were the only familiar thermoplastics during the war, but after it new plastics began to appear commercially at an increasing rate: polystyrene, GRP, polyurethane, polypropylene, polytetrafluoroethylene and the polycarbonates. The 'miracle materials of war', formerly guarded military secrets, became free to benefit society. Factories that had manufactured war materials returned to their original products and applied increased knowledge of materials and processes. For example, a factory previously laminating aircraft parts developed new shapes for plywood chair shells using the newly acquired bonding techniques and synthetic glues. Charles Eames' GRP chairs reflect the fact that he once worked on experimental gliders as part of the war effort.

'The tragedy of America in the Plastic Age.' (*Private Eye*, 1 November 1974)

The Britain Can Make It exhibition organized in 1946 by the Council of Industrial Design at the Victoria and Albert Museum, London, exhibited post-war designs alongside war-time precursors, illustrating positive developments derived from the war experience. But this well-applied war experience was counterbalanced by an over-zealous misapplication of plastics. Widespread application during post-war austerity reinforced the disgraced reputation plastics already had and which the industry is still in the process of setting right.

But the war had already damaged the reputation of plastics in other ways. Developments in Germany in the thirties aroused tremendous jealousy, particularly in Britain. Awareness of the certainty of war had pushed forward German research, and with Hitler encouraging *Heimstoffe* (home-made materials), many substitute materials were developed. For example, in 1935 the Germans were using a synthetic phenolic resin, 'Neoresit', to replace the costly amber resin used for making blood transfusion vessels. They created a product which not only did what was required but possessed the added property of being able to be repeatedly sterilized. During the war the Germans used a vinyl solution, 'Persiton', polyvinyl pyrrolidone, as a substitute for blood plasma. It was obviously very successful as its use later spread throughout Europe and it has been stored for national emergencies in America.

At first the British reaction was to jeer. Many of the German experiments were even considered uneconomic. Allied propaganda labelled their inventions as 'synthetic' whereas in fact they were well-researched substitutes, and it is significant that a German word 'ersatz' was adopted to refer to substitute materials. The trouble is that plastics have nearly always been used as a substitute material, and a high failure rate has contributed to plastics' poor image from the early days.

In the thirties plastics were often used in a manner which the purists amongst us would call 'dishonest'. The flood of replica wooden clocks made of 'Bakelite' conceded nothing to the fact that these cases were totally unnecessary: modern clock movements by that time were mostly constructed with tight dust-proof covers. Concern for the response of materials to this treatment was expressed as early as 1933 by Joseph Sinel who was working at that time for RCA-Victor, Westinghouse and Texaco. He complained that the essential quality of 'Cellophane', its transparency, was often totally obscured by copious over-printing, and to counteract this 'common abuse of Cellophane' he advocated 'rational machine aesthetics'.

The golden era of plastics in the early thirties was followed by a period of crisis and concern at the increasing prejudice against plastics due to the sale of inferior products. Bargain prices bought shoddy goods—faded colours, crazed mugs, cracked boxes and designs that came off in the wash. The whole industry became demeaned and the effects were felt all down the line from designer through producer to consumer.

No one wanted to design products for a second-rate industry. J.P. Gowland writing in *Plastics* in 1938 exclaimed: 'To my knowledge there is only one whole-time plastics designer in the industry who is well paid, but the same man in the printing business would earn twice as much.' Even in November 1975, thirty-seven years on, the editor of *Designer* could ask, 'Why are the services of the professional designer not as much in demand as those of the accountant, the management consultant, the industrial adviser? Why is our design talent under-exploited?'

Caution, even defensiveness, set in in reaction to the poor image of plastics which became common after their usefulness in the war had been forgotten. Post-war failures led to the British Standards Institution laying down standards for plastics, as the Germans and Americans had done years earlier. The setting-up of the Design Centre by the Council for Industrial Design in December 1944 was aimed at promoting design as a vital part of industry. The Council of Industrial Design slide library has a special category in its index for misapplied plastics, labelled 'Plastics, cautionary', and the lessons are being learned, although mistakes still happen.

PVC was particularly accident-prone. A story related by E.W.M. Fawcett at the Historical Discussion Circle on plastics in 1971 recalled the time he was travelling on the New York subway. A lady entered the train wearing a plastic mac. As she sat down it shattered around her. It was so icy outside that her mac had become completely brittle.

Most people still expect plastics to be cheap; they expect a plastic chair to cost less than a timber or metal-framed chair. Yet the raw materials from which plastics are made are no cheaper than any other. Plastics are derived from an expensive fossil fuel, oil, as well as from natural gas and coal, and recently fossil fuels have increased enormously in price. However, against steady inflation throughout the last fifty years, until the sudden recent escalation the comparative cost of plastics had dropped. Even now, although the price of oil has more than quadrupled, plastics are still economically viable, according to a study carried out in 1974 by ICI. The study established that thermoplastics had risen most of all in price because of their direct relationship with oil, but seen against the rise in price of other materials, particularly metals involving very expensive energy-intensive processes, plastics were still competitive. During the paper shortage in 1974 supermarkets were handing out relatively expensive polythene bags in place of paper.

Perhaps the most fascinating aspect of plastics is their mystery and anonymity, that the residue of billions of prehistoric organisms can be transformed, through a series of complex processes, into an endless variety of products. 'Industrial chemistry today rivals alchemy. Base materials are transmuted into marvels of new beauty,' wrote Paul T. Frankl in 1931.

Today these synthetic resins no longer try to simulate precious materials as they once did. Plastics are used so as to declare openly their innate properties as plastics. 'Plastic has climbed down, it is a household material. It is the first magical substance which consents to be prosaic' (Roland Barthes, *Mythologies*, Paladin, London 1973. First published in Paris, 1957). And where the craftsman working with traditional materials such as timber and metal has to do the best he can with what he has, the chemist, processor and designer can together fashion what they will from an infinite combination of chemicals. With god-like power they can create what was once a dream.

The Chemistry of Plastics

MONOMERS AND POLYMERS

Plastics are organic polymers. *Organic* generally describes the chemistry of carbon compounds, materials in which carbon is the chief element. Carbon is the element found in all living things; plants and animals could not exist without it in combination with one or more additional element, such as hydrogen, nitrogen, oxygen and sulphur.

Carbon atoms possess the unique ability to join up to form long chain-like structures of molecules, the structures on which life as we know it is built. All living things are composed of giant molecules. There are some organic materials which are based on mineral, or *inorganic,* substances, for example mica. However, it was the study of the structure of *living* organic materials that laid the foundations of the synthetic plastics industry.

A *polymer* (from the Greek: 'of many parts') is a very long molecular chain constructed from many repeated identical units. The individual units, like building blocks, are called *monomers* (from the Greek: 'a single part'). Monomers are the raw

HOW PLASTICS ARE DERIVED FROM PETROLEUM

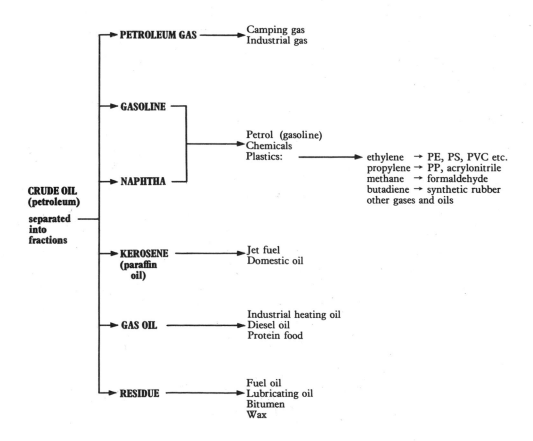

materials from which plastics are built. They are feedstocks from the chemical industry, synthetic chemicals once derived from coal, now produced by the synthesis of the gases distilled from crude oil and natural gas.

Crude oil is a mixture of many different hydrocarbons, molecules composed only of different combinations of hydrogen and carbon atoms. Heating in a refinery separates the hydrocarbons out at different temperatures into their various groups, called 'fractions', namely petroleum gas, gasoline, naphtha, kerosene, gas oil and residual liquids. Each fraction consists of groups of hydrocarbons with similar numbers of carbon atoms. For example, the naphtha fraction consists of hydrocarbons with each molecule containing 6 to 10 carbon atoms with their attached hydrogen atoms.

Most of the fractions are processed further. The naphtha fraction is steam-cracked into the chemicals which form the basis of the plastics industry. At the present time plastics are far more dependent on oil than coal, and on the other important raw material source, natural gas, which converts into basically the same feedstocks.

Carbon atoms can link to hydrogen and to other carbon atoms to form various molecules. The paraffins are simple molecules formed from carbon and hydrogen. Methane (marsh gas), the main constituent in domestic natural gas, is the simplest organic molecule, containing one atom of carbon and four of hydrogen. The carbon atom itself has four valencies (electrical forces, or bonds). Hydrogen has one, nitrogen three, oxygen two. Methane is written like this:

$$CH_4 \text{ or } H-\underset{\displaystyle H}{\overset{\displaystyle H}{C}}-H$$

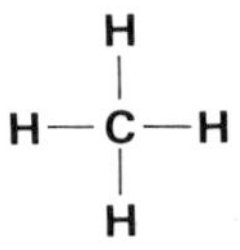

although it is really three-dimensional and more like this:

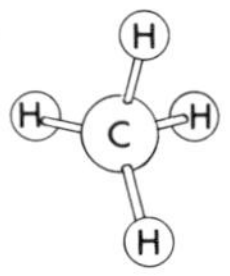

—with the carbon atom in the centre and the four hydrogen atoms bonded around. If two carbon atoms link together in combination with hydrogen they form ethane (C_2H_6):

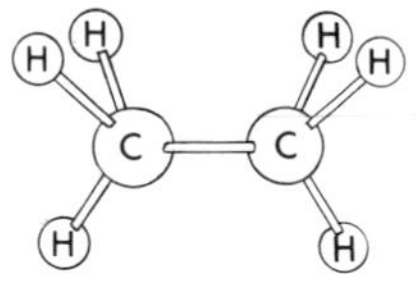

—the sum of the atoms. Three linked produce propane (C_3H_8), four butane (C_4H_{10}), five pentane (C_5H_{12}).

METHANE CH_4	ETHANE C_2H_6	PROPANE C_3H_8	BUTANE C_4H_{10}

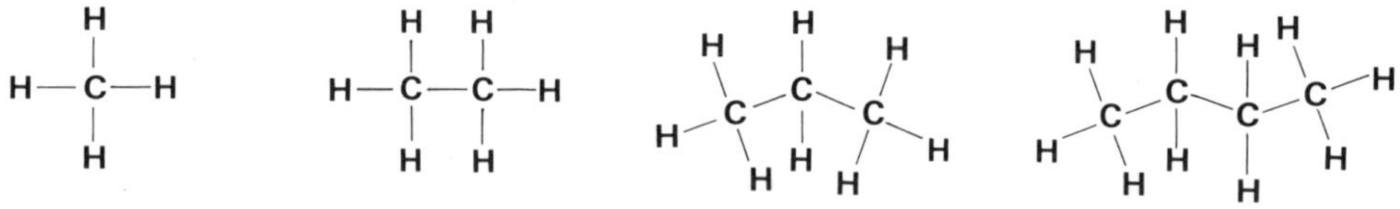

This can be continued to give an immense variety of molecular possibilities.

The olefins are also made from carbon and hydrogen but are known as 'strained' molecules because when viewed in three dimensions the valencies or links are twisted into a position of strain. In ethylene (C_2H_4), two of the carbon valencies satisfy each other and are joined together: $CH_2 = CH_2$ or, diagramatically:

This type of linkage is called an ethylene linkage or double-bond and is not as strong as the straight bond, as in the ethane molecule. It makes the molecule prone to polymerization as it prefers the valencies at right-angles. Ethylene, propylene, butene and pentene are olefin compounds which polymerize to form plastics such as polythene, polypropylene, polyisobutene and poly 4-methylpentene-1.

Polymers are materials with very high molecular weights because they are built up into long and complex chains. For chemical convenience atoms have been assigned particular weights, known as atomic, or molecular weights. For example: hydrogen = 1, carbon = 12. The weight of a molecule is computed by adding together the weights of all the atoms that form it. The longer the chain, the higher the molecular weight. In plastics this information is of use to the chemist in valuing certain characteristics of a plastic, such as the viscosity of the molten plastic which determines the 'melt flow index', the amount of plastics in grammes that can be forced through a standard nozzle at a certain pressure and temperature in ten minutes. It is always helpful to think of plastics composed of giant molecules; the molecular weights of most plastics reach to thousands and sometimes over a million and plastics are referred to as 'high polymers' or 'super polymers'. For example, polyethylene, commonly known as polythene, can have a molecular weight of 560,000. As the number of atoms in a molecule increases, so the compound moves from a gas to a liquid to a solid.

POLYMERIZATION

Polymerization is the process of building up continuous molecular chains from individual identical monomer units. There are two main methods for polymerizing plastics:

1. Addition Polymerization (or linear addition polymerization)
Addition polymerization is the simple linking of polymer units head to tail to form chains by opening out a double bond to free a valency. Linear chains are formed, often with side linkages. The foundations of the process were laid down by I. Ostromislensky around 1912, who analysed the reaction of styrene with styrene. It was this analysis that eventually led to the discovery of many new plastics. One or more different monomers can be linked into the chain—a plastic material produced in this mixed way is called a copolymer. An example of copolymerization is SAN (styrene-acrylonitrile). If three different monomers are linked it is called a terpolymer. ABS is a terpolymer of acrylonitrile, butadiene and styrene. Other addition polymers are: the polyolefins (polythene, polypropylene, polytetrafluoro-ethylene); polyacetals; EVA; polystyrene; the polyvinyls; the acrylics. An example of addition polymerization is the formation of the polythene chain by repetition of the basic ethylene monomer:

If one of the hydrogen atoms in the ethylene molecule is replaced with a different molecule, such as chlorine (—Cl), it results in a different monomer, CH_2CHCl, *vinyl chloride,* represented three-dimensionally as:

On polymerization this becomes PVC:

Similarly, one hydrogen atom replaced by a benzene ring results in the *styrene* molecule:

(Benzene ring)

or if it is replaced by a methyl group, the *propylene* molecule is produced:

2. Condensation Polymerization

In the process of condensation polymerization in order to free valencies for linking up, atoms are removed and a molecule such as water is forced out at each stage in the making of the chain. The polymer produced is thus, in a sense, condensed. Most condensation polymers are copolymers. The condensation polymers are: phenol-formaldehyde; urea-formaldehyde and melamine-formaldehyde; polycarbonates; polyethylene terephthalate; polyurethane resins; and polyamides (nylon).

Polymer chemists have continued to develop new plastics and improve existing plastics, and some amazing properties have been achieved. Like surgeons, the chemists have grafted and copolymerized their compounds.

THERMOPLASTICS AND THERMOSETS

Plastics are materials which under heat and pressure will soften, flow and take on the shape of a mould.

In addition to classification by method of polymerization, plastics can usually be divided into two groups: thermoplastics and thermosets. Plastics made from natural materials, protein and cellulose, can be modified with resins and chemicals to render them either thermoplastic or thermosetting.

Thermoplastic materials are those that can be resoftened and remoulded many times by the application of heat, such as acrylic sheet and polythene pipes. An obvious economic advantage is that they can be recycled. The main disadvantage is that unlike the thermosets, thermoplastic materials are consequently sensitive to heat and will melt. They are usually unreinforced and are often transparent. Before heating, the molecular chains are all tangled up:

When heated and moulded the chains simply slide past each other and retain that position when cooled. If re-heated however, they will slide into another position.

Thermosetting plastics can only be softened and moulded once (hence 'set'), because during the polymerization process the spare valencies of thermosetting plastics link with other atoms in different molecular chains to form permanent chemical bonds. Before heating, the molecular chains are tangled up as in the first illustration above. When heated, the same reaction occurs as in the second illustration. But thermosets need further heating at higher temperatures finally to complete this chemical reaction. This curing process produces a locking effect on the molecules, rather like the knots in a fishing net:

The knots (links) are formed where the chains overlap, producing a rigid three-dimensional structure. This process is called *cross-linking* and is similar to the effect of vulcanization in rubber. A different copolymer may be used for cross-linking.

Because of this chemical reaction, the thermosets are not as flexible as the thermoplastics and can crack and chip on impact, particularly if they do not contain the correct amount of additives and filler. Because of this filler in the mix, which imparts strength, like the stones in cement to make concrete, they are consequently rarely translucent or transparent. Cast phenolics are translucent because they contain no filler.

Cold-setting (i.e. cold-curing) plastics cross-link without the need for applied heat because of the inherent properties of the chemicals involved.

Almost all plastics are now available in different grades – rigid, flexible, self-extinguishing, etc. – controlled by the ingredients in the mixture. Polymer chemistry is like baking; the recipe, a particular combination of ingredients, arranges the atoms in the molecular structure and gives the plastic its particular properties. To confuse matters further, thermoplastic resins have been developed that behave like thermosets, and *vice versa*. Nowadays it is almost impossible to classify a plastic other than by the properties it possesses and the uses to which it is put.

ADDITIVES

The principal material forming the basic plastic mixture is called the resin. Additives are ingredients added to the resin to produce stable plastics. Listed below are the main additives used.

Plasticizer The most important additive, usually a liquid, is a suitable chemical chosen to increase the flexibility of the plastic. The more plasticizer in a resin the more flexible the plastic. The plasticizer also makes it easier for plastics to enter and flow around a mould. Plasticizers are usually organic liquids. The discovery around 1869 that camphor would plasticize bone-hard celluloid and make it into a mouldable material was the first use of a plasticizer.

Filler Also called extenders, fillers are almost always used in thermosetting plastics to make them less brittle and to reinforce their mechanical strength. Thus extended by fillers, less resin is used and a cheaper product can be made without the properties being altered. Exploitation and over-use of fillers, however, results in defective mouldings. Various forms of reinforcement are used: wood flour (softwood sawdust), cotton flock, mica and fibres such as glass and jute. The properties of the filler itself are important: asbestos makes a moulding more heat-resistant; fabric is ideal for curved laminated mouldings; paper for flat laminates.

Stabilizers Some polymers such as PVC are liable to decompose during manufacture, and to counteract this action different types of compounds called stabilizers are added to the mix.

Pigments and Dyes Pigments and dyes are added solely for colour.

Blowing Agents All foamed plastics are basic resins with a blowing agent introduced. Heat, whether self-generated or applied externally, converts the blowing agent into gas bubbles expanding the resin into a foam. If the gas is blown into the resin before it sets, it produces an interconnecting open cellular structure like a sponge. If the agent is included in the compound a chemical reaction produces the gas *in situ* which expands the material into a closed cellular foam, usually rigid.

Catalysts A catalyst helps to start the chemical reaction of polymerization and curing. It can be a chemical itself, or simply applied heat.

Accelerators An accelerator is also a chemical which speeds up the process of curing. It is sometimes referred to as a hardener.

Fire-Retardants Flame-retarding grades of plastics are produced either by the addition of a fire-retardant to the mix or by application afterwards. Growing awareness of fire risk is increasing the importance of these additives, which are becoming increasingly big business.

Anti-Oxidants The effect of oxygen on some plastics during manufacture and use causes degradation. Various chemical anti-oxidants are incorporated to counteract this process.

Antistatic Agents It is now a thing of the past to have cracking, clinging shirts, straight from the washing machine, still grey with dirt attracted from other clothes, or hair standing on end and pulled out by nylon brushes and combs. All this can now be virtually avoided by the addition of antistatic agents.

Natural and Modified Natural Plastics

All plastics can be moulded at some time during processing. The widely held assumption that plastics are recent, synthetic materials obscures the fact that a large variety of naturally occuring plastics has been in use for centuries.

It is known that wax was used to mould seals several thousand years ago. In ancient China, Egypt and Japan various natural tree gums were used for coatings. The Egyptians discovered a multiplicity of functions for bitumen. Amber is a fossil resin crystallized from decayed plants and trees: insects, aeons old, have been found embedded in this natural preservative. When heat-softened, amber can be moulded and small pieces can be compressed together, as for 'Ambroid'. Trade names for moulded amber have included: Ambroid, Beckerite, Gedanite and Glessite. Glass too is thermoplastic and can be re-melted and re-shaped.

SHELLAC *thermoplastic*
Until the development of celluloid in the 1860s, shellac and casein were the most widely used natural plastic resins. Shellac is obtained from lac, a secretion of the *Coccus lacca*, a tropical beetle common in India and Malaya. Encrusted deposits of lac are collected from the trees in which the beetles live, melted and formed into rods, flakes or sheet. More than 2000 years ago the Indians applied it as a varnish on timber, where it formed an excellent coating, protecting the wood from moisture. The insect bodies were used too. Red lacquers were made by mixing resin with the powdered carcasses of the cochineal beetle.

Shellac was also used by the Egyptians and Chinese for sealing and glueing documents. In fact, until the advent of the self-sealing envelope incorporating a rubber-based solution, shellac was a common envelope glue. Shellac was also an ancient decoration for vases and jugs, and a polish.

Nowadays mica laminates are made with shellac. It is still used in the dying art of French polishing, where it is added to wax to strengthen its sealing qualities and to produce a high gloss.

Emile Berliner saw the possibilities of using this dark resin for his nascent sound-recording techniques. In the 1880s he eagerly looked for materials suitable for registering the sounds for the early phonograph. Available materials like hard rubber and wax were either too hard or too soft for prolonged use. He made experimental records of a compound bound with shellac, and later produced the famous 'Berliner' Record label. The record industry thus has the honour of producing the first automatically moulded articles.

As shellac is thermoplastic the mould has to be heated to cause the plastic to flow, and then cooled before ejecting the finished object. Shellac is cheap, and being very easy to mould it is therefore economical, especially as all scrap can be broken up and recycled. It has an excellent ability to reproduce the details of a mould, which accounts for its use for sound records, and all records made before 1930 were moulded from shellac. The resin was mixed with fillers such as cotton flock, then heated and softened, cut into blanks, and pressed in single compression moulds of copper. By 1933 'Bakelite' pressings were made and 'Vinylite' rigid vinyl came into use, more expensive but with improved strength and sound reproduction. When plasticized vinyl took over the role previously held by shellac, long-playing records became a possibility.

Shellac is capable of withstanding extreme changes in temperature, as well as resisting water and oil, and was an economical proposition in the field of electrical insulation. Many electrical insulating components were made of mica bound with shellac, especially during World War II. However, its tendency to warp cut out precision-moulded parts. As it lost its brown tint when purified and could therefore be pigmented, it retained its usefulness as a moulding material for dental plates as well as a host of small everyday mouldings. Today, it is still made into sealing wax.

USA: Cellulak (Continental Diamond Fibre Co.), Lacanite (Consolidated Molded Products)

CASEIN (CS) *thermoplastic and thermosetting*

Casein is the oldest and commercially the most important of the group of natural plastics called protein plastics. The group is derived from the protein materials in animal and vegetable matter and also includes the keratin plastics (bone, horn, hair) and agricultural products like soya beans, corn, oats and wheat.

Casein is made from skim-milk, perhaps not a surprising fact to a cook, familiar with the variety of mouldable solids that can be produced from milk—yoghurts, and cream cheese, junkets, and curds and whey, and the whole gamut of cheeses around the world. But casein is more than just petrified milk. It is obtained by the action of rennet enzyme on skimmed milk, a process that starts off remarkably like that of making junket or yoghurt. The whey is discarded and the curds are cleaned and dried to a powder. The casein is mixed to a dough with water and extruded into rods or sections. Sheets are made up by pressing together chopped or extruded rods and then slicing off layers. When hardened chemically in formaldehyde, it forms into the bone-like material we are familiar with from buttons on old dresses, warm and glossy to the touch. Strictly speaking as it is modified by formaldehyde it is not an entirely natural product.

Casein is known to have been used as a glue since antiquity, but the first patents for the modern material date from 1885 and 1886 in Germany, with German patents released in America at the same time. The patent for 'Galalith' (from the Greek for 'milk' and 'stone') was filed in 1897 by Adolph Spitteler and W. Krische in Upper Bavaria, followed by a patent in Britain in 1889.

Spitteler and Krische were originally searching for a material suitable for making white blackboards but instead stumbled across something with much greater potential. Similar to celluloid it looked like bone but could be moulded to imitate and replace natural bone artefacts.

It was some time before the first commercial production of casein began. To take advantage of a good supply of milk, the Galalith Company was formed as a joint French and German venture. In England casein was known as 'Erinoid', the trade-name of a company which began manufacturing casein from milk supplied by Ireland (hence 'Erin-') in a disused cloth mill at Stroud, Gloucestershire, in 1913-14. One of their first customers were the makers of 'Wimberdar' knitting needles and crochet hooks, Critchley Brothers Ltd. Millions of their needles were made during World War I for the women of England to knit woollies for the Forces. Commercial production began in the USA in 1919. Up to 1940 the main supply of casein for processing came from France; later Britain and the USA established sources in Argentina. It was imported in powder form, just like dried milk, and was moulded by being subjected to moisture, heat and pressure.

The familiar artificial horn and ivory end products were machined from extruded casein profiles. Slices from extruded rod were turned, drilled, filed and carved into millions of buttons and buckles, probably the greatest use along with knitting needles, together with poker chips, umbrellas and bag handles, fountain pens, cigarette holders and boxes, and trimmings of all kinds. Its use spread to paints and distempers, putties, the dressings for paper and cloth, and waterproof coatings for playing cards.

Shellac Union Case lined with velvet. Mid 1860s. 8.4 × 9.5 × 2.2cm (3¼ × 3¾ × ⅞in.). Marked: 'Littlefield Parsons & Co. Are the sold proprietors and only legal manufacturers of Union cases with the embracing riveted hinge. Patented October 14 1856 and April 21 1857.' (Collection and photo Roger Newport)

Used directly from milk, casein is thermoplastic, but moulded under heat and pressure it is cross-linked and sets hard like a thermoset.

Beautiful iridescent or tortoiseshell effects use to be achieved in casein. Several colours extruded together and then sliced up into multi-coloured blanks, or extrusions post-laminated together, imitated mottled or marbled horn. Substitutes for leather, bone, linoleum and horn were produced by the addition of different pigments, or fillers like ivory dust and ground porcelain.

By using the techniques developed for rayon, casein could be converted into artificial silk. It was forced through a spinneret and the filaments cured in a hardening solution. However, the resulting yarn, only available in America under the name of 'Aralac', was so weak that it could only be used in conjunction with other materials like wool or viscose rayon.

Casein is still used for making decorative buttons. However, it did not seem to have the moulding potential which the industry had hoped for. 'Communications on the subject in the technical press are rare' complained a German author in 1921, trying to find a place for casein in industry.

When the use of inflammable celluloid was widespread during the 1920s, casein was tried as a safer substitute but proved impractical, owing to its high moisture content. Cascin is hygroscopic and warps and cracks, as casein buttons often do.

UK: Erinoid (Erinoid Ltd), Keronyx (Aberdeen Combworks Co.), Lactoid (BX Plastics)
USA: Ameroid (American Plastics Corporation), Aralac (yarn, National Dairy Products Corporation), Casco (glue, Casein Co.), Galorn (George Morrell Corporation), Karolith (American Plastics Corporation), Kyloid (Kyloid Co.)
GERMANY and FRANCE: Galalith (International Galalith)

KERATIN
Human and animal bodies are made up of certain polymer materials. Their properties have been known and used for years. Keratin is the protein that forms hair and fur, wool, feathers, nails, hoof, horn and bone. It has a long-chain molecular structure. The permanent wave is a plastic moulding, exploiting the thermoplastic properties of keratin in the human hair. Glues are made by the extraction of gelatin from fish and animal bones or with dried blood.

Bone, ivory and tortoiseshell are natural keratin-based polymers and can be modified and moulded. Horn and tortoiseshell can be softened in boiling water, pressed flat and moulded into shapes such as buttons. Thicker slabs can be built up by laminating layers together under heat and pressure, as was popular in the eighteenth century. Even two centuries later a great many plastics on the market still attempt to imitate tortoiseshell and horn. As a cheap substitute for glass, horn could be pressed so thinly that it became transparent. The Smithsonian Institution in Washington possesses a lanthorn dated 1740 when keratin began to replace glass.

SOYA BEAN PLASTICS *thermosetting*
Protein is found in very large amounts in agricultural cereal products, particularly corn, oats, flax, hemp seeds and beans. In the thirties and forties there was much experimentation with every possible material from green coffee berries to scrap leather. But the only crop to achieve development potential was the soya bean. Even today, concern for our depleting natural resources continually focuses on soya beans to supplement the materials that sustain us.

The soya bean consists of approximately 20 per cent oil, 40 per cent protein, 30 per cent carbohydrate and 10 per cent fibre and ash. The protein extract can be powdered, mixed with phenolic resin and cross-linked under heat and pressure to form moulded shapes. The chief developer and exploiter of the soya bean was Henry Ford who began experimenting late in 1931 and subsequently set up a special research laboratory. He had previously laid waste to thousands of bushels of

Henry Ford swinging an axe against the rear panel of his two-door sedan to illustrate the strength of his new Rogers sheet material (soya-bean reinforced phenolic sheet), 1945. (Ford Motor Co.)

watermelons, carrots, cabbages and onions, macerating them in his search for an agricultural plastic from which he could mould an entire car.

His dreams were remarkably prophetic: 'No matter what we may guess as to the proportion of automobile parts that can be built from the fruit of the field our guess will fall far short of the eventual result.' His persistence was rewarded, for although it was not possible to press out an entire car, it was possible to produce many smaller components.

In 1940 he had ready the complete set of dies for moulding these parts and October 1941 saw the public unveiling of the first plastic motor car. Cheaper and lighter than steel, its impact strength was ten times greater.

The new car body was made of fourteen plastic panels fitted to a tubular steel frame, with windows and windshield of plastic 'safety glass'. The panels of soya protein fibre were hardened with phenol-formaldehyde resin and formed in a press. Heat and pressure thermoset the panels into their unalterable shape. Ford used the salvaged oil from the soya bean in the enamel on his cars.

Although the vehicle design in the illustration does not look very different from the standard model of that time, a landmark in plastics history had been reached, and hope had been implanted into America's wartime agricultural industry. Time magazine commented, 'If the dream is true, the technological novelty known as plastics has graduated from its celluloid and 'Beetleware' phase into an instrument of industrial revolution.'

The material itself did not ultimately meet design hopes. Like casein it tended to absorb water, but it did pave the way for GRP and other, later plastic car bodies.

FURFURAL, ZEIN AND LIGNIN PLASTICS *thermosetting*

Laboratory experiments were carried out with other regenerated fibres as well as soya beans. Farm waste of all kinds was processed in the search for cheap moulding materials: straw from wheat, rye, oats, corn, flax, barley and rice; seed hulls from oats, rice and cotton; corn cobs, sugar cane, tobacco and peanut stalks. A handful of trade names survived commercially: 'Vicara' fibre from sweet corn cobs; 'Ardil' fibre from peanut shells; 'Oil-stop' (Irvington Varnish), a cashew nutshell plastic; and even a coffee-bean plastic, 'Caffelite' (Caffelite Corp., USA), from beans imported from Brazil. Almost all the experimentation took place in America, mainly during World War II.

Furfural (from the latin *furfur*, meaning bran) is a liquid produced by the distilling of oat hulls, corn cobs and much of the waste produce listed above. It is a very cheap source of materials. The distillation, furfur-aldehyde, reacts with formaldehyde, or is condensed to produce the furfural resins.

'Durite' (Durite Plastics Inc.) was one American trade-name for phenol-furfural goods which were compression- and transfer-moulded into mainly electrical items such as small magneto heads for aircraft and thermostats, but its main use was in paints.

Zein is the protein found in the gluten in corn, and is processed like soya beans into a meal mixed with thermosetting resins. During the American depression some companies, particularly Quaker Oats, sensibly attempted to utilize their farm waste in this way.

Lignin is a natural wood resin akin to the phenolic-type resin. Wood contains approximately 25 per cent lignin. It is separated from wood pulp, paper waste and sawdust and used to enrich the lignin already present in wood cellulose. It was used in the moulding of thin sheets in the early thirties, laminated into panels for the building industry. The sheets were black and so had to be coated or painted. Hardboard is the modern sheet version made from pulped wood cellulose.

Like furfural, lignin was used extensively in World War II.

NATURAL AND SYNTHETIC RUBBER

Rubber, whether natural or synthetic, differs from other plastics in its basic elasticity, its rubberiness, its elastomeric quality. It has a whole complex technology of its own, much older in fact than that of our modern plastics industry.

Natural rubber has a long-chain molecular structure and can be softened, moulded and cured like other polymers.

Goodyear's discovery of the vulcanization process in 1839 not only made moulded rubber products possible but became an important part of plastics technology. All the equipment used in the vulcanizing and moulding of rubber was later adapted for the natural plastics such as shellac and gutta percha, and the early synthetics. Although there is a natural tendency to think of rubber in its simpler forms, hard rubber looks and feels like certain other plastics and it is sometimes difficult to tell the difference.

Finally, many of the newer polymers are classed as synthetic rubbers, and most plastics today can be manufactured in elastomeric grades, for example, the thick moulded soles of our fashionable clogs and wedge shoes.

Natural Rubber (NR) *thermosetting*

Latex (Caoutchouc) is the white milky sap collected from the rubber tree *Hevea Brasiliensis*, originally a native of Brazil. Latex is a natural linear polymer with a long-chain molecular structure and consists of 35 per cent rubber (isoprene, $CH_2 = CCHCH_3 = CH_2$, a monomer of carbon and hydrogen), 60 per cent water and 5 per cent resins and other substances. To get rid of the high water content acid is added which makes the emulsion coagulate, separating the water from the solids. The solids are then rolled into sheets, corrugated like knitting to allow the substance to dry more quickly, and then cured like ham in a smoke house, to enable it to be shipped in a preserved state. This process of eliminating water is now being superseded by the use of extruders. After further processing the raw material is calendered into sheets, from which shoe soles for example can be stamped, extruded into products like inner tubes, and compression-moulded with heat into motor-tyres, hot-water bottles and other everyday items.

Rubber had been known for centuries in South America. Columbus discovered the South American Indians playing with amazing bouncing balls, and their games are depicted in old Inca drawings; RAPRA (The Rubber and Plastics Research Association) and the Science Museum, London, each possess one such ball, two to three hundred years old, looking rather like a golf ball with the outer cover taken off. RAPRA also possesses a rubber boot made by the Indians in the nineteenth century as a copy of European leather boots they had observed. Somewhat like a galosh it has convincing stitching patterns moulded onto the surface. Liquid latex can be either dip-coated, spread as a coating onto some base fabric, or expanded into soft foam. The dipping process, wherein a former is simply dipped into liquid latex, is the most ancient of all the techniques. The South American Indians are believed to have dipped their feet into latex for instant shoes.

Rubber only become familiar in Europe around 1770, when Joseph Priestley, an English chemist, discovered its graphite-erasing properties and erasers were put on the market. At the same time in France rubber was made into a mouldable paste by dissolving latex in turpentine and ether. The British scientist Michael Faraday was the first to isolate the isoprene molecule in 1826, although it was not given that name until 1860.

In 1820 Thomas Hancock discovered that rubber became a soft dough if it was cut up by spiked rollers, heated and kneaded. His famous masticating machine can be seen at the Rubber and Plastics Research Association. Toughened by the addition of a filler such as carbon black, it was discovered that rubber could be moulded like plastics. But there were problems, as it was easily deformed. In cold temperatures it became uselessly hard, while in hot weather uselessly sticky. One of the most famous applications of natural rubber latex was the invention in 1823 by Charles Macintosh of a process for manufacturing rainproof fabric by sandwiching the tacky gum between layers of cotton—the macintosh.

Made from rubber latex, soft foam was invented in England in December 1928 by the Dunlop Rubber Company Research Centre, who developed it by beating latex with liquid soap in a home foodmixer and baking it in the oven like a meringue. This is still one of the two basic methods used for moulding foam rubber. The milky latex is frothed up with chemicals and poured into a mould and the foam is cured in an oven. After washing and drying it ends up in its familiar form: cavity cushions (ultra-soft white foam slabs penetrated by regular holes) or shaped toys such as 'Bendy Toys' sprayed with washable paints.

A wide variety of shapes and sections can be moulded, in many different grades of firmness, from the softness of cotton wool to the hard densities used for such purposes as public transport seating. Its excellent shock absorption makes it suitable for all kinds of upholstery. Cushions are usually built up by hand from different densities, with a firm core bonded to an outer surface layer of very soft foam.

The density of the foam depends on the ratio of foam to air and the size of the cavities within the block. In foam rubber these air cells interconnect.

The second method for making latex foam is the 'Freeze-gel' process (the Dunlopillo SP process) which was developed by Talalay. The creamed foam only partly fills the mould, which is then closed and a vacuum applied to expand the foam. It is frozen solid by carbon dioxide blown through it to make it gel. The foam is finally cured at a higher temperature and then washed and dried. Foam rubber, whether natural or synthetic, is non-allergic and will not harbour dust mites or bacteria. It is thus an ideal material for asthma or allergy sufferers and for hospital equipment, including specially shaped beds, splints, foam rings and burn dressings.

UK: Dunlopillo SP Foam (Dunlop), Texfoam, Vitafoam.

Non-foamed Natural Rubber
The wide range of familiar uses of natural rubber include gloves, wellington boots,

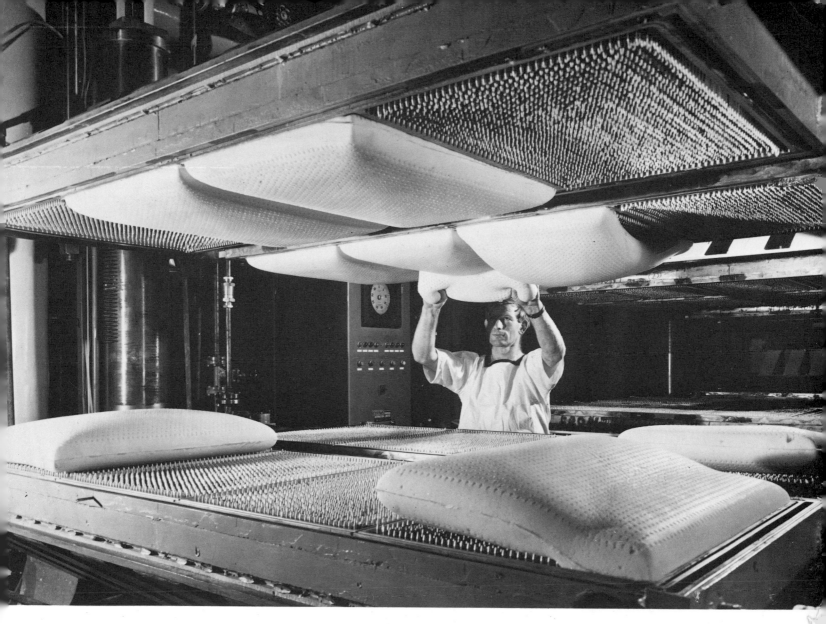

balloons, babies' bottle teats, frogmen's suits, very large flexible containers, tool handles, fabric coatings, and tubing. Balloons are made by dipping a metal former into the latex, building up several layers, and stripping the mould after curing (vulcanization).

A special form of isoprene was developed to mould the tyres of the Transporter driven by the Apollo 14 mission, the first wheels to leave tracks on the moon.

Vulcanization, Hard Rubber, Ebonite, Vulcanite

In 1839 Charles Goodyear in America discovered the vulcanizing process for rubber. It was also developed at the same time in England by Thomas Hancock, and a British patent was filed later in 1856, five years after Goodyear had exhibited his Hard Rubber Goods at the Crystal Palace in 1851.

Goodyear found that by adding sulphur to masticated rubber and heating it, he obtained a material that could neither be reheated nor re-moulded. Even more important was his discovery that the rubber had become exceedingly resilient—in fact, rubbery.

By prolonging the vulcanizing process he could vary the degree of hardness, and eventually obtained a very hard form, called hard rubber in America, or ebonite and vulcanite in England.

Vulcanization has the same effect as curing or cross-linking thermosetting resins, and the molecular pattern obtained is very similar.

Latex foam pillows being removed from their moulds at the Dunlopillo factory at Pannal, Harrogate, 1975. (Dunlop Ltd)

25

Crude rubber: when pulled:

The chains are tangled up, but free to slide past each other.

Vulcanized rubber,
with bridges of sulphur: when pulled:

The three-dimensional network is immovable.

When vulcanized, the isoprene molecule has become a polyisoprene molecule, like a plastics polymer.

Vulcanization made rubber available at last as a mouldable material, with thousands of potential uses. Solid rubber tyres were exhibited at the Great Exhibition of 1851, having been patented first by R.W. Thomson. Their use was later taken up during the development of the bicycle. A direct result of the discovery of vulcanization was the invention of the first pneumatic tyre in 1888 by a veterinary surgeon from Belfast who developed it for his son's bicycle. His name was John Boyd Dunlop. That invention was to have a long-lasting effect on the subsequent design and production of the motor car.

Many materials can be vulcanized with sulphur. In the thirties even the giant seaweeds growing plentifully along the Pacific coast of the USA were vulcanized into 'Sea Caoutchouc'!

The development of hard rubber was an historical landmark. Unlike shellac and bitumen, vulcanized hard rubber does not change shape when re-heated, and was thus the first thermosetting material and the first natural polymer to be chemically altered by man.

When the vulcanizing process is taken to its limit, the intense heat finally makes the rubber very hard. Hard rubber tubes, sheets and rods were stock items, while other shapes could be moulded on demand. Its good insulating properties and its chemical inertness made it a popular material until first celluloid and later phenolic resins replaced it for certain uses.

Unaffected by chemicals it was used for containers for storage batteries, acid measures, cisterns and buckets, and especially in the manufacture of photographic trays and tanks where it remained in use for almost a hundred years. It found many other uses, in particular buttons, jewelry, combs, telephone receivers, cutlery handles, and smokers' pipes. All kinds of wheels were moulded, from toy wheels to steering wheels to caster wheels up to ten inches in diameter with metal insert bearings. Being naturally black like bitumen, colour was limited. But mottled colours could be achieved with pigments and fillers.

Hard rubber mouldings were usually compression-moulded in simple two-part moulds. Hollow articles could even be blow-moulded, while hard rubber sheet could be drilled, scored, stamped or used for lamination.

A major use for vulcanite was for moulding dental plates, gum facings and teeth fillings (the latter were first patented in America in 1916). They replaced celluloid carvings tinted with pigments which 'very quickly disappeared leaving the elephantine portion as unmistakable evidence of the sculptor's art' (*Plastics* 1937).

UK: Bulwark (Redfern's Rubber Works)
USA: Ace (American Hard Rubber Co.), Endurance (American Hard Rubber Co.), Luzerne (Luzerne Rubber)
FRANCE: Gallia

Synthetic Rubber (SR)

Synthetic elastomers are a complex technological field, extensively documented elsewhere. They are man-made materials possessing the stretchy property of natural rubber, and can be vulcanized. Synthetic rubbers imitate the long-chain molecular structure of natural polymers, and are built up from copolymerized synthetic monomers, such as butadiene with styrene, or butadiene with acrylonitrile.

Until World War I the growing need for rubber in motor car production was satisfied by increasing the production of natural rubber in the Malayan plantations. The advent of war greatly increased the overall demand for all natural resources; and due to the blockade of the supply of rubber to Germany from the East, I.G. Farbenindustrie at Leverkusen was the first to develop a small amount of synthetic rubber—methyl rubber—made with alcohol.

By the beginning of World War II the picture had changed enormously. Progress in polymer chemistry and petroleum distillation had made it possible to develop synthetic rubber from oil, and ways had been found of copolymerizing different rubbers from a variety of monomers, in the same way that different plastics are produced. Germany now had Buna (SBR) rubbers, made by the polymerization of butadiene and styrene, begun by Bayer in 1929; a version was produced later in America by a different method, Buna-S, when the German patents became available. By 1932 Du Pont had made neoprene (GR; Government Rubber) commercially available under the trade-name 'Duprene'. This was a very strong and durable copolymer rubber made from the monomer chloroprene ($CH_2 = CHCCl = CH_2$). Neoprene is now put to all kinds of uses, from water-proofing fabric for divers' suits, to coating re-usuable inflatable sheets for building concrete vaults. In 1934 the first American automobile tyre had been made completely from man-made rubber by the Dayton Rubber Manufacturing Company. The German scientists had copolymerized butadiene with acrylonitrile, producing nitrile rubber (NBR), while an isobutylene isoprene copolymer was developed known as butyl rubber ('Butyl'). Polysulphide rubbers, such as 'Thiokol', first made in 1928, were developed in America, and by 1945 enormous bladders of 'Thiokol Mareng' were used to convey oil.

Following Pearl Harbor and American involvement in World War II in 1941, America's source of rubber from the East Indies was effectively severed, and as about 90 per cent of the world's supply of natural rubber comes along that route this quickly provided a stimulant to the young synthetic rubber industry. By the mid-1960s the output of synthetic rubber had overtaken that of natural rubber.

One of the most important synthetic rubbers today is styrene-butadiene rubber (SBR) which is extensively used in the production of motor car tyres. An unusual use of SBR is as an emulsion sprayed along the Norfolk coast by Dunlop to arrest erosion and promote vegetation.

Synthetic rubber production has grown tremendously in the form of soft flexible foam, found widely in homes and in industry. Popular references to 'foam rubber' usually mean foams of the polyurethane type, and are either polyester (AU) or polyether (EU). In 1950 Dunlop set up a pilot plant to produce its own foamed rubber from synthetic polyurethane latex, under the trade-names of 'Dunlopreme' and 'Dulon'. New types of synthetic rubber are continually appearing, including ethylene-propylene rubber (EPR), polybutadiene (BR) and polyisoprene, the artificial 'natural' rubber.

Synthetic rubber is incidentally a constituent in chewing gum, replacing chicle, a natural thermoplastic rubber latex, as the 'gummy' base.

UK: Breon and Butaclor (BP Chemicals), Cariflex (Shell Chemical Co.), Dunlopillo, Dulon and Dunlopreme (Dunlop)
USA: Butaprene (Firestone Tire and Rubber Co.), Butyl Rubber (Standard Oil Co.), Chemigum (Goodyear Tire and Rubber Co.), Hycar (Hycar Chemical Co.), Hypalon (Du Pont), Neoprene and Duprene (Du Pont), Nextel (3M), Thiokol (Thiokol Corporation).
Elastomeric fibres: Lycra (Du Pont), Vyrene (US Rubber Co.)
GERMANY: Baypren (Bayer), Buna-S and Buna-N (I.G. Farbenindustrie)

Silicone Semi-Organic Polymers (SI) *thermoplastic and thermosetting*

Silicon is one of the most plentiful elements on our planet. It is the element found in sand, combined with oxygen to form what is usually called silica (silicon dioxide, SiO_2). For a depth of twenty-five miles it forms one quarter of the earth's surface. When the tough, inert properties of inorganic silicon are joined with the polymer-forming organic element, carbon, the silicone compounds produced have remarkable properties.

Silicones are manufactured in numerous forms, from hydraulic and lubricating fluids, liquid resin coatings and adhesives, to elastomeric mouldings varying from very soft and gel-like to rigid and solid. The process of vulcanization of some of these mouldings needs applied heat, others vulcanize at room temperature. Medical silicone rubbers are often reinforced with powdered silica. The only soft material for making human implants is silicone rubber. Other implant materials like 'Teflon' (PTFE) and 'Dacron' (polyester) are basically rigid, whether moulded into solid blocks or fabric.

The pioneering work on silicones was done in England by Professor F.S. Kipping at Nottingham University in the early years of this century. Like many organic chemists at that time, he was looking for a crystalline mouldable resin. Instead he found what he described as 'uninviting glue'. It was not till the mid-thirties that an American, Dr J. Franklin Hyde, developed silicone commercially as an insulating lacquer. But the war stimulated progress. In 1943 the Dow Corning Corporation was established at Midland, Michigan, by the Corning Glass Works and the Dow Chemical Company, with the specific purpose of developing silicone. During the war the corporation was the sole supplier of silicone fluids and grease for military use.

A major step in the direction of silicones' ultimate use in implants was the discovery that if the insides of glass receptacles and tubing were coated with silicone, blood did not clot and water did not cling to the surfaces. Hence its use now as an inner coating for medical equipment such as bottles. The first actual silicone rubber implant was carried out by Dow Corning in 1950 in replacing a urethra tube, which was a very premature use of the material since very little information was available on the tolerance of the human body to silicone rubber. Most data referred to silicone fluids. The first implant had no adverse effects and requests from research physicians for silicone rubber medical parts began to assail Dow Corning.

In response to this challenge, a new medical silicone rubber, 'Silastic S-9711', was compounded by Silas A. Braley in 1953. In 1955 the first commercial implant was designed, the Hydrocephalus Shunt, to drain spinal fluid from the brain. This particular unit has saved the lives of thousands of children. A simple and clever design feature is the incorporation of extra long tubing, which, once implanted, uncoils as the child grows.

Progress was so rapid that in 1959 Dow Corning built a special unit to concentrate on the medical application of silicone rubber. One of their famous developments is the breast implant which in 1962 replaced pads made of synthetic sponge (expanded polyethylene or polyvinyl alcohol-formaldehyde). A seamless, transparent sac is made of elastomeric silicone rubber gel, with a piece of woven 'Dacron' cloth attached to the back. This fabric acts as a trellis allowing the natural tissue to interweave with it so that the implant becomes one with the body. Early artificial breasts appear stiff and hard, like old-fashioned brassières, compared to the latest almost shapeless models, which sag naturally. The first Dow Corning breast implant

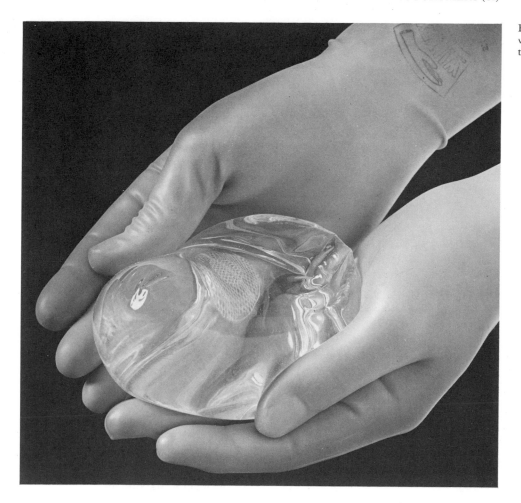

Breast implant of silicone rubber gel with woven 'Dacron' polyester cloth attached to the back. (Dow Corning Corp.)

is still in place. The injection of silicone to form breasts as opposed to silicone implants is frowned upon by the medical profession.

Other implants made of silicone rubber include ears, skull plates and jaw bones. Silicone rubber joints straighten arthritic fingers and remobilize wrists, arms and big toes; bile and tear ducts can be made to refunction; and testicle and penile implants can revitalize masculine egos. How far are we from the reality of plastics muscles or kidneys—the Plastics Man?

Many plastics have been tried for artificial hearts, and as yet there is no perfect material. The heart valves in the first pacemaker, designed by Dr William H. Chardack, were moulded in silicone, which is now used mainly to insulate the pacemaker wiring. Actual hearts have been moulded in silicone reinforced with 'Dacron' cloth.

Silicone is probably most familiar domestically in the form of water-repellent furniture polishes, coatings, paints and caulking sealants. Self-curing silicone rubber sealants for interior or exterior building joints are estimated to last thirty years before becoming brittle.

Anti-fungal properties can be added to give extra resistance against the mildew found in hot damp kitchens and bathrooms. Sealants are also made from many other plastics such as polyurethane and acrylic for areas where plaster tends to shrink and crack. 'Breathing' fabrics impregnated with silicone are made into raincoats. The silicone does not form a complete film over the fabric but waterproofs the fibres and repells rainwater while allowing air to pass through the material. Injection of silicone into brick walls fills cavities so as to form a waterproof barrier which dampness cannot penetrate.

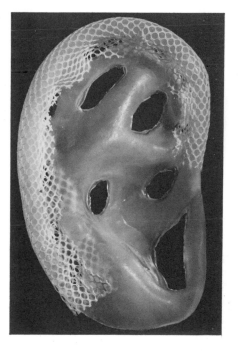

Silicone rubber ear armature with 'Dacron' polyester cloth and holes for tissue to grow through. (Dow Corning Corp.)

29

Plastic Man (Original published in *Police* No. 1, National publications Inc., 1939, 1940, 1941, 1942.) 'Rubber Man' would be a more appropriate name for him.

Like PTFE, silicone rubber is applied as a coating to baking-pan surfaces. Silicone rubber chocolate and biscuit moulds in factories save on washing and regreasing. Silicone rubber is even sprayed as a release agent onto moulds for plastics. Flexible moulds are easily peeled off rigid low-pressure structural foam components, and the silicone is particularly good at reproducing minute details, such as those on ancient fossils and prehistoric tools, or the timber graining on structural foam furniture.

GUTTA PERCHA *thermoplastic and thermosetting*

Gutta percha is a material of the past, overtaken by advances in new plastics. Most people look blank on hearing this mysterious name, although older golf players may recall 'gutties'—golf balls of solid gutta percha. Shellac, rubber and gutta percha were the first polymers to be moulded in sheet and tube during the second half of the nineteenth century.

Gutta percha was an early natural substitute for *papier mâché*, leather, wood, paper and metals. Often thought of as a rubber-like material, it is not like an elastomer at all, although like natural rubber, it is a polymer of isoprene with the same formula C_5H_8. It is neither a runny tree sap nor rubbery when processed. It is a hard dark material that has to be stripped by hand from beneath the bark of the palaquium tree in the jungles of Malaya, Borneo and Sumatra. It can also be obtained from the twigs and leaves pruned from those trees. Softened by heat it becomes plastic and mouldable and on cooling returns to its former hardness.

The material was first made known to Europe in 1843 in the form of implements and sculptures made by Malayan natives. A unique collection of such objects belonging to the Gutta Percha Company (now Telcon Plastics Ltd), the first gutta percha processor in Britain, was unfortunately destroyed by a bomb in 1940. Stripping the palaquium trees and destroying the branches and leaves was a very extravagant process. Exploiters cut down whole trees at an increasing speed, and threatened the very livelihood of the plantations. The discovery of a similar plastic gum called balata extracted from the *Mimosups balata* trees of South America eased the problem, as it could fulfil some of the functions of gutta percha. Still of significance, balata is also a hard natural gum and a thermoplastic that softens on being heated, sets on cooling and can be resoftened.

Around 1915 the Selborne Plantation Co. was formed and established a plantation which succeeded in producing more than 100 tons a year without destroying the trees by simply collecting gutta percha from pruned branches. The material was shipped to London in yellow brick-sized blocks stamped 'SP' (Selborne Plantation). At the Gutta Percha Co. factory, the blocks of crude gutta percha were cut up and taken into a boiling room or vat room to be softened in hot water and mechanically kneaded. Passing on into a rolling room, the material was fed through rubber rollers like laundry, which compressed it into sheets. At that time gutta percha sheet was produced by rollers onto a continuous belt. To make thin sheet, women were initially employed to stretch the pliable sheet by hand to a thickness of a few thousandths of an inch as it emerged from the rollers. The next step was to pin the sheet to the sides of a conveyer belt where it cooled.

Mouldings were made by compressing hot gutta percha in cold moulds. Early moulds were made of lead, often copper-coated and supported in an iron outer case. Sometimes gutta percha and rubber mixtures were vulcanized to produce a harder product.

The invention of electric telegraphy and the founding in 1846 of the first Electric Telegraph Company was fundamental for the Gutta Percha Company. Previously cables had been coated with tar, wax and pitch until gutta percha revolutionized the process, its first-class electrical properties having been determined by Michael Faraday in 1843. In 1850 the first submarine telegraph cable in the world was laid from Dover to Calais. The Gutta Percha Company made the first batch of 'twenty-five nautical miles of No. 14 Birmingham gauge copper wire covered with great care in gutta percha to half-an-inch diameter'. The machinery at that time could only make the cable up in a variety of shorter lengths; these were transported down the Thames by steam-tug to Dover where they were joined up into one coil.

One end of the cable was laid out from Cap Griz Nez lighthouse at Calais and simply ran over the rocks and into the sea. The other end 'ran from a horse-box in Dover Station yard, across the road, along a wooden stage which was there in connection with the building of the Admiralty Pier, and into the sea'. On 28 August the Dover cable was joined to the Calais cable on board the steam-tug in the middle of the Dover Straits. Every sixteenth of a nautical mile a block of lead was attached to the cable to weigh it down. After the joint had been made, the tug continued the journey to Cap Griz Nez and set up the print-out instrument and the first telegraph message, addressed to Louis Napoleon, was received in France. Unfortunately the first historic message and its return signals were 'so jumbled that no sense could be made'. Retardation due to induction had mixed up the letters, which should have been transmitted at a slower pace. A second attempt with modified apparatus was a great success, and this was followed in 1856 by the first step in connecting America with England, which was completed ten years later.

The resin in gutta percha and balata reduces their electrical insulating properties, while the protein in rubber retains water. By combining protein-free rubber with resin-free balata, the Bell Laboratories in America produced an improved insulating plastic, patented as 'Paragutta', which helped to make the dream of the transatlantic telephone cable possible. When polythene was discovered by ICI in 1933, all these problems were solved.

Most of the gutta percha products in the Great Exhibition of 1851 at the Crystal Palace, London, which earned the Gutta Percha Company a medal, would be of great interest to collectors of antiques nowadays. Many items actually looked like the bronzes they were copying: busts of Homer, Grecian-style inkstands, pen trays, royal emblems, plaques and coats of arms.

Jug and chemical bottles moulded by hand in gutta percha by the Telegraph Construction and Maintenance Co., London. (John Topham Picture Library)

above:
Samples of 'Erinoid' casein mouldings made
by Erinoid Ltd between 1929 and 1940.
(BP Chemicals International Ltd)

Three mouldings in 'Parkesine', 1861-1868,
by Alexander Parkes; box lid with moulded
hinges and two small religious book covers
inlaid with metal and mother-of-pearl.
(Plastics and Rubber Institute, London.
Photo Chris Smith)

The chemical resistance of gutta percha was ideal for early photographic Talbotype trays, collodion baths and daguerreotype equipment, all compression-moulded. Other mouldings appear surprisingly modern, showing that consumer demand has not changed. Shaped moulded lengths of gutta percha are exactly the same as those which are marketed now in timber or the new expanded plastics foam: a solid frieze to fix round a funeral coach, frames, crests with cupids, mouldings for churches and pulpits, and emblems for courts of justice.

The uses to which gutta percha was put were as varied as can be imagined, both practical and fantastic—from medals and buttons to miners' helmets, from clock cases and bands for machinery to ice skates and tubing for primitive 'intercom' systems. In those days a patent was obtained easily, encouraging the use of gutta percha and leading to some lively times—perhaps too lively, as very few of these items survive today.

The only purpose for which gutta percha and more specifically balata still reign supreme is golf balls, although recent developments threaten even this last stronghold. Original golf balls of the 1850s were made of solid gutta percha, moulded into smooth spheres. When it was later discovered that a damaged surface gave more control to the flight of the shot, the surface of the bronze mould was designed with the familiar 'bramble' marking, a pattern integral with the ball. As the gutta percha was dark in colour, the balls were coated by rolling them between hands covered in white paint. At the turn of the century the moulding room of the Gutta Percha Company was producing 100,000 solid gutta percha balls a week.

The solid gutta percha ball gave way to the rubber-wound core, one hundred and five feet of vulcanized rubber thread densely wrapped round a liquid centre. This core was placed between two shells of compression-moulded gutta percha and both sections were compression-moulded together. The vulcanized surface was finally painted and a trade-name was embossed on the ball.

Telcon still processes gutta percha and balata straight from the jungle, which nearly all now goes to cover golf balls, though a little can be found in Chatterton's Compound and some in dental fillings.

INORGANIC PLASTICS

Certain inorganic natural materials have plastic properties. Although they are not organic carbon compounds, clay, glass, mica, wax, cement and bitumen are all nonetheless mouldable.

Bitumen *thermoplastic and thermosetting*

Of these inorganic yet mouldable materials bitumen has the closest characteristics to the organic plastics. Needing only to be melted and cooled until it hardens, bitumen could be termed thermoplastic. In 1947 N.J.L. Megson in his British Council publication *Plastics* described bitumen as a typical plastic.

Bitumen compounds such as asphalt, coal-tar, gilsonite and pitch are natural waste products from the distillation of petroleum. They also occur naturally as deposits in Cuba, California, Texas and in the famous Pitch Lake in Trinidad. The Egyptians employed bitumen thousands of years ago for surfacing roads and caulking boats. Now we are more familiar with its use in mouldings for battery boxes, old fashioned bottle stoppers, handles for pots and pans, electrical goods and switches. Its application is limited by its dark colour.

Bitumen mouldings are made by pressing softened material into cold moulds (i.e. cold moulding).

Bituminous moulding compounds were discovered in 1909 after celluloid and shellac. Emile Hemming was trying to improve upon the properties of shellac, and used one of the asphalt bitumen compounds to bind asbestos wool in a 3:1 mixture of asbestos to asphalt. With the heat resistance of asbestos the moulding could be cured in an oven after being formed under pressure.

The moulding took anything up to three days to dry out, which would now be uneconomical, but as a consequence was very thoroughly cured, even tougher than some of the thermosetting plastics then available. When mouldable 'Bakelite', the first synthetic organic polymer, made its debut in 1909, it invaded all areas of design but could not replace bitumen for storage batteries. Bitumen's resistance to acid still ensures its use today, along with polypropylene.

'Ebonestos' was the trade name for bitumen mouldings marketed by Bernard Weaver's company, Ebonestos Limited, founded in 1900.

Glass-Bonded Mica

Other inorganic mouldable materials such as clay, porcelain and glass have been used for centuries as hand-shaped containers, especially for food. The high instance of breakage is far outweighed by the material's non-toxic properties, its low cost and ease of manufacture.

Glass-bonded mica can be considered one of the first ceramoplastics. Developed in Britain in 1919 it was widely used in World War II because of its dimensional stability. It was the only material suitable for high-frequency electrical components and the same properties make it useful in the computer industry today. Mica dust compounded into an inorganic thermosetting resin was moulded into many components for radios, such as valve holders and coil formers, under the trade-name 'Mycalex'. Mica splittings bonded with phenol-formaldehyde could resist temperatures up to 537°C. By the late thirties mica was being incorporated into melamine formaldehyde and as 'Micarta' was moulded and laminated.

UK: Micanite (The Micanite and Insulators Co.), Mykalex (Mykalex Co.)
USA: Mykalex (Mykalex Corporation and General Electric Co.)

CELLULOSE AND CELLULOSE PLASTICS

Celluloid and all cellulosic plastics are derived from natural cellulose, the least complicated of all polymer materials. Cellulose is the cheapest and most abundant raw material in the world since all plant cells are composed of it—wood, straw, cotton, vegetables, flax, sugar cane.

DERIVATIVES FROM WOOD CELLULOSE

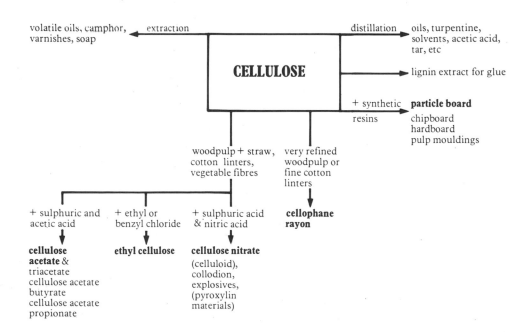

Celluloid 'ivory' and 'pearl' mouldings, 1890s-1940s, including boxes by Halex Ltd in 'Xylonite', and cylindrical cotton wool and hair-pin boxes with blow-moulded lids. (Photo Fritz Curzon)

opposite above:
'Carvacraft' Writing Desk Equipment, 1948-51, manufactured by John Dickinson and Co. Ltd., including: double inkstand, blotter, desk diary holder, (*right*) spring-loaded telephone index holder (*left*), paper-knife, and 'Propelling, Repelling and Expelling' vulcanite pencil with rubber and lead-holder, advertising The Staveley Coal & Iron Co. Honey-mottle cast phenolic resin supplied by Catalin Ltd. After World War II the Dickinson presses that had compression-moulded papier-mâché and 'Bakelite' shell casings were used to mould phenolic and urea 'Gadeware' products, designed by Charles E. Boyton, with distinctive trade-marks designed by Harry J. Earland. The 'Carvacraft' range consisted of twenty-three accessories for the desks of the 'Executive, Director, Board Room and Conference Room'. (Author's collection. Photo Fritz Curzon)

opposite below
Celluloid 'tortoiseshell' mouldings, 1890s-1920s, including powder bowl, cigarette case, glove stretcher, various boxes and tray made from 'Xylonite' sheet and rod. (Photo Fritz Curzon)

When burnt, a cellulose-based plastic object gives off the smell of burnt paper. Being bio-degradable is a major advantage of cellulose products—they simply rot and return to the earth. Natural cotton wool is one of the purest forms of natural cellulose: the long fibres are used for spinning and textiles, the short unused fibres, the cotton linters, are used as the basis for cellulosic plastics.

Cellulose has a long-chain molecular structure but without treatment it is not capable of being moulded under heat and pressure. That ability was only developed after 1891, when Cross and Bevan discovered the viscose process and the fact that the cellulose chain could be modified in a variety of ways.

All cellulose products are a regenerated form of natural cellulose, whether very refined wood pulp or crude chips and particles. With various treatments or bonded with synthetic resins they are transformed into celluloid, 'Cellophane', rayon, acetate, paper, cardboard, particle board and pulp mouldings.

Cellulose Nitrate (CN)

Cellulose nitrate was first prepared in 1833 by Henri Bracconet, the director of the Botanical Gardens at Nancy in France, when he mixed cellulose in the form of sawdust and linen with nitric acid. Twelve years later at Basle University, Professor Christian Schönbein repeated the experiment with ordinary paper made from wood cellulose treated with nitric acid. He named the highly inflammable transparent material that resulted cellulose nitrate, commonly but incorrectly referred to as nitro-cellulose, and he patented it as an explosive. It was the beginning of the development of the first man-made naturally based plastic, the first modified natural polymer—celluloid.

In the following years it was discovered that cellulose nitrate was itself soluble in a mixture of alcohol and ether, producing a solution that became known as collodion. Although exceedingly inflammable, its property of shrinking as it dried was used in medicine to seal small wounds. The ether evaporated rapidly leaving a thin plastic film. It had similar uses in early photography. In 1851 F. Scott Archer developed the 'wet plate' process whereby a film of collodion gelled into a thin skin on a glass plate. When the cellulose nitrate solvent had evaporated, it left a thin flexible emulsion which could be stripped off the glass plate. It was light-sensitive and capable of holding an image. Dyed with pigments, collodion solution was sold as a ladies' nail varnish.

The more nitrogen cellulose nitrate contains, the more inflammable it becomes. A 12 to 13 per cent nitrogen content is known as gun cotton (cellulose trinitrate), a well-known explosive. Collodion and cellulose nitrate lacquers (pyroxylin) contain 11 to 12 per cent. In celluloid, which appeared later, the nitrogen content was reduced to 10 to 11 per cent. However, celluloid is still combustible, and even in modern factories cigarettes and anything to do with fire are totally banned. Many of the early celluloid mouldings were lethal and would not be allowed on the market now. The stories are legion of combustible gentleman's collars: 'But woe to the man who carelessly dropped a glowing cigar or pipe ash on his collar! There would be a flash of flame and the collarless victim ended with a badly burned neck.' The celluloid dolls familiar in the forties and early fifties are quite illegal now.

Parkesine

Although America is credited with the successful patenting of celluloid, the early form of it, Parkesine, was produced by an Englishman, the Birmingham inventor Alexander Parkes. He took cellulose nitrate in the form of cotton fibre or wood flour dissolved in nitric and sulphuric acids, and mixed it with vegetable oils such as castor oil and wood naphtha. The combination made a dough which could simulate ivory and horn, and could be textured and painted. He fully realized the potential of his discovery and first exhibited a few moulded household goods—knife handles, combs, plaques, and medallions—at the 1862 International Exhibition in London where they aroused sufficient interest for him to receive a bronze medal.

Parkesine was shaped by pressing the heat-softened dough into moulds or carving by hand, then it was painted or inlaid. It was far cheaper to produce than costly gutta percha and leather, or even rubber, for which demand was rapidly growing. At last a material had appeared with tremendous potential, but in need of further development. In a paper of 1865 Parkes had listed all the functions that were later to be fulfilled by celluloid and cellulose acetate: 'Spinners' rolls and bosses . . . combs, brush backs, shoe soles . . . whips, walking sticks . . . buttons, brooches, buckles, pierced and inlaid work, book-binding, tubes, chemical taps and pipes . . . waterproof fabrics, sheets and other articles for surgical purposes.' He succeeded in making an enormous range of mouldings including sheet, rod and tube; pieces of treated cloth and rigid impregnated lace; counters, varnished materials; a variety of balls, especially billiard balls, formerly made of ivory; umbrella and knife handles; buckles, clasps, buttons and beads; mirror backs moulded in classical relief; and plaques and large circular medallions. His knife handles really look and feel like ivory. When not copying other materials Parkes achieved a very attractive brightly coloured mottled effect by mixing coloured Parkesine—reds, greens, blues, oranges—in a manner that was later echoed by phenolic resins. Now over one hundred years old, these exhibits, kept mainly in the Plastics and Rubber Institute in London, are the oldest plastics in the world, but the colours are still as bright as ever.

Unfortunately Parkes' company, the Parkesine Company, was short-lived. Parkes encountered many problems with his product and seemed to lose interest in his discovery. The company finally foundered in 1868 and was taken over by his friend Daniel Spill. Spill was soon manufacturing the new improved 'Celluloid', following the filing of a patent by the Hyatt Brothers in the USA in 1869.

Plaque moulded in 'Parkesine', 1861-1868, by Alexander Parkes. (Science Museum, London)

39

Celluloid (CN) *thermoplastic*

Parkes' moulding dough dried very hard and bone-like, nearer to ceramics than plastics. The Hyatt Brothers in America discovered that the reaction of cellulose nitrate with different chemicals from those used by Parkes resulted in a far more plastic material.

The missing link was camphor, which Parkes had used as a solvent for dissolving his pyroxylin without realizing its significance. The Hyatts nurtured the process and successfully launched 'Celluloid' into production. It was described in *Landmarks of the Plastics Industry* (ICI, London 1962) as 'the first truly domestic plastic'.

The name 'celluloid' with a small 'c' has become the generic name for all cellulose nitrate plastics. It is strictly speaking only the American trade-name. The English equivalents were 'Ivoride' and 'Xylonite', so named on account of its derivation from xylene (wood naphtha). Other American trade-names were 'Pyralin', 'Viscoloid', 'Fiberloid' and 'Nixonoid', the Greek suffix '-oid' implying a shaping material.

The Hyatts were encouraged in their work on cellulose nitrate by an offer of 10,000 dollars from Phelan and Collender in Albany, New York, manufacturers of billiard balls, who were badly in need of a substitute for ivory. A large reward for a seemingly small beginning presaged an even greater significance. John Wesley Hyatt, like Parkes, was a man of inventions. He obtained 250 wide-ranging patents, the 'Celluloid patents', starting in 1869. In 1870 the Hyatts set up the Albany Dental Plate Company, the first firm to produce celluloid commercially, replacing dark-coloured hard rubber dentures with an illusion of real gums. In 1871 it became the Celluloid Manufacturing Company; in 1927 it merged with the Celanese Corporation to form the Celanese Celluloid Corporation, now the Celanese Corporation.

Celluloid began to catch on, and other companies were formed. The Fiberloid Company was established in 1894, to be purchased much later by Monsanto. By 1915 one of the largest producers of celluloid in the USA was the Merchant's Manufacturing Company, later bought by E.I. du Pont de Nemours and Company (Du Pont). The Viscoloid Company was formed in 1901 to mould combs from pyroxylin. Du Pont bought Pyralin in 1915, and Viscoloid in 1925, which together became the nucleus of their present plastics section, and 'Viscoloid' was adopted as a Du Pont trade-name.

Part of the credit for the Hyatts' success must go to their engineer, Charles Burroughs. He was the man behind the machinery, designing the tools for the company. The modern plastics industry is indebted to the processes and machinery that Burroughs developed for processing cellulose nitrate, which, as J. Harry DuBois points out in *Plastics History—USA,* are as important as the material itself, and many of his methods are still in use today.

In England, Daniel Spill had taken over the Parkesine Company, renaming it the Xylonite Company to produce 'Xylonite', but he too, like Parkes, had had problems with celluloid production and went bankrupt in 1874. The following year the persistent Mr Spill created Daniel Spill and Company and moved to a new site. In 1877 L.P. Merriam joined with Spill and three others to form the British Xylonite Company, which enjoyed great success in making celluloid collars and cuffs. It was eventually taken over by BIP Ltd in 1973.

Daniel Spill unfortunately became embroiled in a quarrel with Hyatt over patent infringement, which finally destroyed him. It took several years of disputes in the American courts to see who really did discover the plasticizing of pyroxylin. Spill sold up all his investments in the company, contracted cyrrhosis, dropsy and diabetes, and died in 1897.

The Manufacture of Celluloid

The advantages of celluloid sheet are that it moulds easily at low temperatures and can be cured at room temperature. Consequently none of the later machinery developed for moulding plastics at high temperature and pressures is necessary. However, as already described, cellulose nitrate is highly combustible and unstable

opposite
The Wurlitzer Automatic Phonograph, Model 600, 1938, made from 'Catalin' cast phenolic resin in what the Catalin Corporation called the 'modern interpretation'. (*Modern Plastics*)

A block of 'Xylonite' (Celluloid) being
planed into sheets. (BIP Ltd)

and great care must be exercised in its manufacture.

It is also easy to machine celluloid to precise dimensions, and to tint it in a wide
range of colours and tones from 'water white' to dense black. Metallic pigments,
producing iridescent effects, are familiar in celluloid, which is considered by many to
be the most beautiful of all the plastics.

Various pasta-like processes are undertaken in the making of celluloid. Cellulose
nitrate and, now, synthetic camphor are first dissolved in a solution of nitric acid in
order to make a dough. The dough is pressed and filtered and formed into lengths
like spaghetti. As John Merriam pointed out in an article in the BIP *Beetle Bulletin* 36,
Autumn 1975, 90 per cent of the 'Xylonite' process is aimed at extracting all the
combustible nitrate solvent through evaporation. The 'spaghetti' is then pressed into
thick pastry-like sheets, called 'hides', which are laid on top of one another in a pile
15 centimetres ($5\frac{7}{8}$in.) high. Laminated together under heat and pressure they merge
into one solid yet flexible block. The solvent content has by now been reduced to 15
per cent. The blocks are planed into sheets of varying thicknesses and hung up like
animal skins to 'cure'. Any residual solvent dries off, and any stresses that have built
up in the manufacture are lost as the material expands and contracts. A final pressing
irons out any wrinkles and gives the desired matt or polished surface finish. The
sheet can then be machined, glued, hot-pressed, or blow-moulded to form dolls,
table tennis balls and other hollow items. Cellulose nitrate sheet shrinks around
wood as it cools. This property was used to make the first plastic piano keys, by a
process that was still in use in the fifties, and was also exploited for shrinking around
wooden blocks to form ladies' shoe heels. Extruded celluloid is made by chopping
'hides' into small pieces (nibs) which are fed into a screw extruder, heated and forced
out through a die.

The height of celluloid's popularity was reached between 1900 and 1920 in the

USA, and between 1920 and the mid-thirties in Britain. The three main visual effects of celluloid mouldings were tortoiseshell, ivory and pearl.

Celluloid has been so effective as a substitute for tortoiseshell, that even now antique dealers frequently make mistaken identifications. Celluloid 'tortoiseshell', appropriately enough called 'shells', is made in various shades of translucent brown. As John Merriam explained in his article, the creative technique and experience for making these shells have been handed down in the British Xylonite factory by word of mouth, and the process is often more the result of intuition than a standard chemical recipe. The basic method involves compressing different coloured sheets together, planing the blocks back into sheets and re-pressing them in a different order. It can be seen why the combinations are endless.

Mock tortoiseshell invaded every corner of the domestic scene. Dressing tables in particular were covered: trays and powder boxes, brushes and manicure sets, ear trumpets and spectacle frames. Most of these uses have now been superseded by non-exploding plastics, although guitar plectrums are still produced in imitation tortoiseshell cellulose nitrate.

Ivory celluloid was used for the same purposes as 'tortoiseshell' and can be more easily recognized by its faint stripes.

The most attractive grades in celluloid sheet are the iridescent shimmering effects called 'pearls', often seen on the backs of hairbrushes, mirrors and powder boxes. Particles of lead phosphate incorporated into the dough create crystal-like angular and symmetrical patterns. Although celluloid is the only plastic in which these particular effects are possible, the handling of it is probably a dying art.

The chemical and water resistance of celluloid has also been exploited for fountain pens, set squares, protractors (formerly of tin and brass), and slide rules, replacing the bone handles of drawing equipment, which tended to split and stain with ink.

Over a quarter of a million of these 'cels' were needed for Walt Disney's *Snow White*, 1939. These were drawings on 'Pyralin' celluloid film. (*Modern Plastics*)

Cellulose nitrate was even used as a coating. It transformed cotton into imitation leather and was popular for book bindings and the manufacture of gold and silver evening shoes. In 1851 *Harpers'* offered delighted testimonials supporting collars, cuffs and 'ventilated bosoms', swearing that celluloid effected a permanent cure for sore throats and lung disease sufferers.

Probably the most famous use of celluloid, however, was in the form of transparent film. Eastman Transparent Film was first patented in December 1889, and its use as cinema film contributed to the creative development of motion pictures. But it was so inflammable, and fires broke out so frequently, that insurance companies would often insist on fire-proof projection booths. Fortunately it was later replaced by non-flammable cellulose sheets which were known as 'cels'. Although celluloid has long been replaced by triacetate film, the transparent flexible sheets used for animation are still called 'cels' and often mistakenly referred to as celluloid. When celluloid film replaced the glass plates formerly used in photography, it influenced the development of the hand-held camera.

The first shatterproof 'safety glass' was made from celluloid, which was laminated between two layers of glass or used alone. It is remembered because its low resistance to ultra-violet rays made it fade and turn brown. During World War I the British Xylonite Company manufactured forty and a half million eye pieces for gas masks. The Plastics and Rubber Institute collection contains a pair of World War II goggles made of celluloid with khaki fittings—for a horse. It is as large as a bikini top, an early Emanuelle Khanh plastic bra.

Today almost half of BIP's output is for table tennis balls and, despite experiments with other materials, it looks as if cellulose nitrate is going to hold the monopoly in that field for a very long time to come. Still used for dice, guitar plectrums and drum trim, it is also used for mortar bomb casings because of its combustibility.

Celluloid as an Artistic Medium

Celluloid was the first plastic to be used by artists since the ancient use of lac and bitumen. For some reason, artists did not use rubber until the late 1960s, with the exception of the gutta percha sculptures of the Malayan natives. However, Parkes directly reproduced the classical styles of bronze and wood in Parkesine.

The model for Naum Gabo's *Woman's Head* is the first recorded use of plastics in sculpture. The head is bolted together from opaque sheets of celluloid, chosen because it was easy to construct models closely simulating the sheet metal of the final work. *Woman's Head* is an early constructivist work, growing from its corner like a collage in relief. It predates the *Realist Manifesto* that Gabo wrote in 1920, and which his elder brother Antoine Pevsner co-signed, a manifesto that formulated a new concept of abstract art, revaluing time, space and matter. In its championship of an art dedicated to the modern age, to technology and its materials, it has influenced art and design up to contemporary times.

Transparent celluloid sheet provided a medium for the Gabo brothers' ideas as well as symbolizing a new technology. It was ideally suited to constructing forms from their reduced abstractions, from the inside out so to speak, and enabling one to experience the inherent structure rather than simply the outward form.

The form of *Torso* by Antoine Pevsner is defined through the interpenetration of celluloid and copper planes. The repeated use of curved and flat planes held in mathematical tension is typical of constructivist work. Working against the background of revolutionary Russia and the revitalized function of the artist in society, Gabo also used other plastics mediums as they appeared on the market: casein, cellulose acetate sheet, and acrylic sheet.

UK: Cascelloid (Cascelloid Ltd), Halex (Halex Ltd.), Xylonite (British Xylonite Co.)
USA: Amerith (Celanese Plastics Corporation), Dumold (Du Pont), Fiberloid (Monsanto), Herculoid (Hercules Powder Co.) Nixonoid (Nixon Nitration Works), Protecto (Celluloid Corp.), Pyralin (Du Pont, now a trade-name for their polyimide resin), Viscoloid (Du Pont)

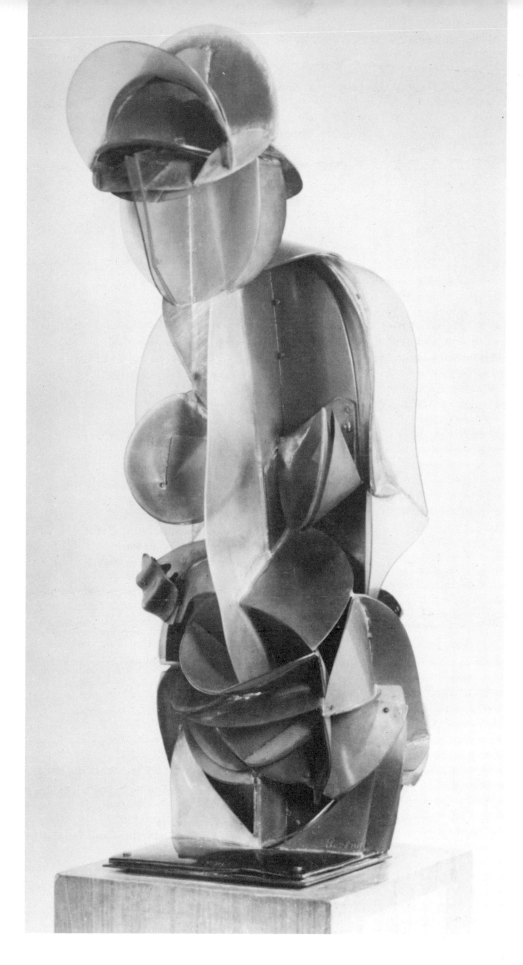

Antoine Pevsner, *Torso*, 1924-26. Celluloid and copper, 73.6cm. (29in.) high. (Museum of Modern Art, New York)

Naum Gabo, *Woman's Head*, 1916-17.
Celluloid and metal, 62.2 × 48.8cm
(24½ × 19¼in.). (Museum of Modern Art,
New York)

Cellulose Acetate (CA) *thermoplastic*
For a while celluloid was the only plastic material available for moulding the goods
that Alexander Parkes had foreseen. All attempts to make celluloid non-flammable
were unsuccessful and chemists searched for a similar material without the hazards
of combustibility. Non-flammable cellulose acetate had been known for almost as
long as cellulose nitrate (it had been made experimentally in the late 1860s by Paul
Schützenberger), but it only became commercially viable in 1911, after Charles
Cross, Edward Bevan and Clayton Beadle developed the viscose process, a new way
of making the already known viscous solution of pure natural cellulose into a
mouldable non-synthetic plastic fibre for spinning, as fine and soft as real silk. Their
process followed the work of Count Hilaire de Chardonnet, who in 1884 first
produced artificial silk commercially by spinning threads of collodion made from
nitrated cotton.

The viscose process starts with cellulose from wood or any other natural plant made into a viscous solution and filtered. It is then 'ripened' to a controlled degree and forced through metal spinning dies like mincemeat, to emerge as filaments which are then cleaned, bleached, hardened and dried. The final yarn is made by spinning these filaments together. The name 'viscose' is considered old-fashioned and the yarn is now referred to as 'rayon', although 'viscose' is often found on knitwear labels. In 1904 Samuel Courtaulds and Company secured the British rights to manufacture rayon by the process, producing in particular large amounts of black crêpe mourning clothes. 'Tufcel' and 'Vincel' fibres by Courtaulds are even more soft and absorbent than the cotton they replace.

Cellulose acetate's non-flammability derives from acetylation, the treatment of cellulose with acetic acid instead of nitric acid. It actually smells of vinegar when it burns.

In 1894 Charles Cross and Edward Bevan achieved the complete acetylation of cellulose to produce triacetate, soluble in chloroform, and in 1903 this triacetate (primary cellulose acetate) was treated to form a secondary acetate (diacetate) soluble in acetone. This was the cellulose acetate used as 'dope' on aircraft and later by the Dreyfus brothers to make 'Celanese' fibre. In 1911 in Switzerland Camille and Henri Dreyfus at last improved Cross and Bevan's earlier viscose process and produced acetylated cellulose economically. The result was the first commercially available cellulose acetate, in the form of a liquid for coating. It was profusely used as dope during World War I for coating the fabric-covered wings and fuselages of aircraft—at last a non-flammable resin. Up to 1918 it was only available in this form.

Cellulose acetate dope had been in such demand that immense stocks had been built up and a use had to be found after the war ended. Experiments with the surplus stocks finally led to the production of cellulose acetate sheet, rod and tube, which immediately took a large share of the celluloid market, and to the first acetate rayon yarn, 'Celanese', made in England at a factory established in Derbyshire by the Dreyfus brothers.

Like rayon, the acetylated cellulose acetate dope is spun through spinnerets and can be toughened ('oriented') by stretching. It is made into underwear, bathing costumes and silky 'Dicel' shirts and sweaters. The acetylation reaction was continued to its furthest point in cellulose triacetate (TPS). In fibre form it is highly crease-resistant, moth-proof and water-repellent. As it is thermoplastic it can be permanently pleated. Triacetate is made into all kinds of drip-dry clothing.

After 1918 cellulose acetate rod, sheet and tube came into production, but sheet in particular found an important place as a non-flammable substitute for glass. Celluloid may have been the first shatterproof 'safety glass', but cellulose acetate added the dimension of non-flammability to achieve even greater safety.

In the early thirties British Celanese brought out 'Celastoid' sheet laminated between two layers of glass, to introduce a shatterproof flexible pane that did not turn brown. 'Blast-proof' windows were made with slices of cellulose acetate reinforced with wire or string netting. Cellulose acetate film took over movie and photographic film, was made into goggles and watch covers, and particularly found use in the aircraft and automobile industries for windows, cockpit covers, gun turrets, bomber noses, domes and 'blisters', and the new contact lenses, all later replaced by acrylic. It was not so good as a heat-resistant material for electrical insulation.

Moulded Cellulose Acetate

It was not until 1929 that cellulose acetate appeared in powder form as the first thermoplastic suitable for injection moulding. It was first developed in Germany, and the United States and the rest of Europe could not injection-mould cellulose acetate powder until the mid-thirties. For quite some time it was the only injection-mouldable thermoplastic available. Any colour or combination of colours was possible, transparent, translucent or pearlescent, with different grades of hardness depending on the plasticizer used.

Cellulose acetate is available in three forms. First, as granules and powder made from broken-up sheet, used for mouldings, or stock extrusions, produced by the thermoplastic processes of extrusion, injection moulding, rotational moulding, compression moulding and thermoforming. Secondly, sliced off a moulded block, like celluloid, it is available as both rigid and flexible sheet. Thirdly, it can be cast as a film for use in the packaging industry.

Cellulose acetate tool handles were far better than wooden ones which tended to split along the grain. Golf tees were made from cellulose acetate in the forties and fifties, and millions of spectacle frames were cut from sheet. Also, being thermoplastic, tube could be softened in hot water and drawn by hand over metal tubing to make smooth covers for perambulator handles, or solvent-shrunk round hand-rails on omnibuses. In the twenties and thirties flexible cellulose acetate tubes replaced glass for medical containers—the familiar tubes for creams and lotions, with moulded plastic caps.

Moulded cellulose acetate brought colour into the home: the colourful 'Tefra Refillable Toothbrush' moulded in 'Lumarith' in 1931 had removable bristles—a progressive design at a time when people still bathed in white tin tubs! Coloured plastics began to appear too in the automobile industry, as did steering wheels, handles, knobs, plates, and fascias in a variety of mottled colours, some even simulating natural stone in moulded 'Lumarith'. In particular, cellulose acetate's

Desk monophone, Model 300, designed by Henry Dreyfuss, 1936. Moulded in Tenite cellulose acetate by The Tennessee Eastman Corp. (*Modern Plastics*)

qualities of translucency and transparency were widely applied in lighting and it was often embossed to imitate parchment.

The allied fields of packaging and display exploited the new cellulose acetate film materials, and developed transparent, see-all containers of all shapes and sizes. Clear 'windows' appeared in cardboard boxes. Thermoformed 'blister-packs' made a debut. Now cellulose acetate film has a wide variety of uses including record sleeves, laminated book jackets and cosmetic packaging, as well as boxes for cakes, flowers and toys.

The Blue Ribbon Tower, also called the Helicoidal Tower, at the Paris International Exhibition of 1938 was eighty metres high and constructed from sky-blue transparent corrugated 'Rhodoid' sheet. Illuminated at night, it appeared as a symbol and token to the future. It stood out on the Paris horizon like a second Eiffel Tower, symbolizing both man's architectural and his engineering expertise as well as France's merchant navy.

Part of a range of lampshades made from 'Rhodoid' cellulose acetate sheet, 1937. (*Plastics*, Temple Press)

UK: Armourbex (BX Plastics), Bexoid (BX Plastics), Celastoid, Clarifoil and Cellastine (British Celanese), Cellomold (F. A. Hughes and Co.), Cellophane (British Cellophane), Cinemoid (cellulose acetate for cinema and theatre, British Celanese), Darelle (Courtaulds fibre), Dorcasite (Charles Horner Ltd), Doverite (Dover Ltd), Erinofort (Erinoid Ltd), Isoflex (BX Plastics), Kodapak (Kodak), Novellon (British Celanese), Rhodoid (M & B Plastics), Tricel, Tufcel and Vincel (Courtaulds), Wireweld (cellulose acetate sheet reinforced with wire, British Celanese)
USA: Arnell (triacetate fibre, Celanese Corporation of America), Bakelite Cellulose Acetate (Celanese Celluloid Corporation), Fibestos (Monsanto Chemical Co.), Fortisan (fibre, Celanese Corporation), Lumapane and Lumarith (Celanese Corporation), Nixonite (Nixon Nitration Works), Plastacele (Du Pont), Tenite I (Tennessee Eastman Corporation, now Eastman Kodak Co.), Vimlite (Celanese Plastics Corporation)
FRANCE: Rhodoid (Rhône-Poulenc)
GERMANY: Cellidor (Farbenfabriken Bayer), Cellon (Dynamit-Nobel AG)

Cellulose Acetate Butyrate (CAB) *thermoplastic*

Cellulose acetate butyrate, commonly referred to as butyrate, or acetobutyrate, was an American development of the early forties. CAB is a cellulose ester, produced by the action on cellulose linters of acetic and butyric acids and their anhydrides. A tougher type of cellulose acetate with improved weather- and heat-resistance, and an excellent glossy surface finish, it is used for tool handles, steering wheel covers, pen barrels, toys and road signs. A more recent application is for oil and natural gas pipelines.

It was very popular in the States in the forties for weaving shiny belts, braids, handbags and braces, and widely used for public transport upholstery, woven basket-chairs, tables and screens, and to coat yarns, a method developed in France where it was adopted by high fashion. CAB 'Plexon' yarns were based on either cotton or rayon, and were moulded in various sections, such as round, flattened or even square. Very beautiful rainbow-like effects could be created by using a tinted resin over coloured thread.

A particular use for CAB was for moulding musical instruments and reeds. It did not freeze the lips nor need warming up, and was extremely light. Musicians seemed to be satisfied with its tonal qualities, although it was well known for its rather unpleasant plastic smell.

UK: Cabulite (M & B Plastics)
USA: Plexon (yarns, Freyberg Bros.—Strauss, Inc.), Rexenite and Rexonite (yarns, The Rexenite Co.), Rextrude (Rextrude Co.), Tenite Butyrate (Tennessee Eastman Corporation)
GERMANY: Cellidor Bsp. (Bayer)

Cellulose Acetate Propionate (CAP or CP) *thermoplastic*

Cellulose acetate propionate is also a cellulose ester, but in this instance treated with acetic and propionic acids and their anhydrides. It is similar to cellulose acetate butyrate, but generally tougher and used on a smaller scale, for pens, pencils and formerly telephone handsets

USA: Forticel (Celanese Corporation)
GERMANY: Cellidor CP (Bayer)

Ethyl Cellulose (EC) *thermoplastic*

Ethyl cellulose was first developed in Germany around 1912 but not produced there commercially until 1935. It results from treating cellulose first with caustic soda and then with ethyl or benzyl chloride. It was used during the war as a protective coating for machine parts in transit, which up until that time had been shipped heavily coated with grease and then wrapped by hand. This lengthy process was replaced by simply dipping the parts into hot resin which rapidly shrank as it dried into a very tough tight skin, similar to the present-day shrink-wrapping process. It was also used as a moulding material and as an adhesive. Being non-toxic and colourless it is very useful in food packing.

UK: BX Ethyl Cellulose (BX Plastics), M & B Ethyl Cellulose (M & B Plastics)
USA: Ethofoil (Dow Chemical Co.), Hercules Ethyl Cellulose Flake (Hercules Powder Co.), Lumarith Ethyl Cellulose (Celanese Plastics Corporation), Methocel (Dow Chemical Co.), Nixon Ethyl Cellulose (Nixon Nitration Works)

Regenerated Cellulose Fibre Products

Among other kinds of cellulose fibres that can be moulded, either used naturally or treated with synthetic resins, are 'Cellophane', particle board and moulded wood pulp products.

Cellophane (a Du Pont trade-name, but now a generic term) is made from regenerated wood pulp and is basically viscose rayon in film form. It is an inflammable product, and on burning the distinctive wood or paper smell of

The Blue Ribbon Tower ('Helicoidal Tower'), Paris International Exhibition, 1937. Blue 'Rhodoid' cellulose acetate sheet, 80 metres (262ft) high. (*Livre d'Or*, Royal Society of Arts, London)

opposite
The *Plia* folding chair designed by Giancarlo Piretti in 1969 for Anonima Castelli, Bologna. Piretti's first design in plastics, thermo-formed in Bayer's 'Cellidor', a toughened version of cellulose acetate, superseding the original in acrylic.

Us Army bugle, 1945, weighing only 10 oz., injection-moulded in cellulose acetate butyrate by the Elmer E. Mills Corp., Chicago, in conjunction with the Chicago Musical Instrument Co. (Tennessee Eastman Corp.)

cellulosic plastics is apparent. The viscose solution is extruded in sheet form through a narrow slit and treated with the same chemical processes as rayon filaments. It gets its final polish by passing between calender rollers, and can be lacquered to make it moisture-proof for wrapping food. 'Cellophane' revolutionized the packaging industry.

The most familiar type of particle board is chipboard, made from softwood particles. Similar boards are manufactured from a wide variety of other residual wastes such as sugar cane (for bagasse board), flax and straw. Most of these boards are bonded with urea-formaldehyde resin in a dry process. Besides flat sheets, chipboard can be moulded into simple shapes, used in the past for schoolroom and children's furniture, although now not so much in evidence.

Pulp mouldings are made by a wet process by which wood fibres are pulped together with water-repellent synthetic resins and binders, usually phenol-formaldehyde. A preform or 'felt' is moulded to the rough shape of the end-product over a wire former, by pressure or vacuum. The preform is dried and then pressed in matched moulds, to squeeze out the water and to give the final detailed shape and finish.

Hardboard is made in this way, also usually strengthened with added phenol-formaldehyde synthetic resin, although the natural lignin content of the wood is actually sufficient to bond the sheet.

Pulp moulding is a cheap process for making high impact mouldings. Developed about ten years ago, its full potential has not yet been fully exploited, particularly for furniture. The mouldings are light-weight, thin-walled but strong; they are corrosion- and impact-resistant, and have been used for the backs of TV and radio cabinets, for luggage, record player cases, office equipment, and many automobile accessories—fascias, boot liners, gear box covers, glove compartments, filters and rear shelves.

One of the earliest forms of pulp moulding was a type of woodware patented as 'Bois Durci' by Charles Lepage in France and England in 1855. Sawdust made from a hardwood such as rosewood or ebony was soaked in egg or blood albumen diluted with water. Dried to a powder, it could be formed in a steel mould under heat and pressure.

'Prestfibre', 'Pimfibre' and 'Preform' are trade-names of British Moulded Fibre Limited.

Synthetic Plastics

PHENOL-FORMALDEHYDE (PF) *thermosetting*

Up to now the plastics described have been either naturally occurring materials such as shellac, gutta percha, rubber, viscose; or natural materials that have been modified by simple chemicals, such as casein, vulcanized rubber, Parkesine, celluloid and cellulose acetate. The latter can be called semi-synthetic as the chemical reactions involve no complete transformation of the natural ingredients.

In 1907 the first totally synthetic material was developed in America—phenol-formaldehyde resin, often abbreviated to phenolic resin, but commonly known by the trade-name 'Bakelite'. 'Bakelite' was patented by Leo Baekeland and later manufactured in America by the Union Carbide Company, who later used the trade-name to encompass many of their other resins such as styrene, vinyl and the polyolefins. Generally 'Bakelite' now refers to Union Carbide phenolic resins only, unless otherwise indicated (for example, Bakelite Urea, or Bakelite Cellulose Acetate).

Dr Leo Baekeland was one of the greatest researchers in the history of plastics, and many modern synthetics derive from his work. As a young professor of chemistry he went to America from Belgium and made a fortune with the invention of 'Velox' gaslight photographic paper, the first paper on which photographs could be printed in artificial light. Around 1899 he sold his company to Eastman Kodak and built his own laboratory in Yonkers, a northern suburb of New York, where he proceeded to resolve the mystery of the phenolic combination.

Formaldehyde had first been polymerized by Butlerov in 1859, and by the early 1870s Adolf von Baeyer had reported that phenols and aldehydes combined to form horn-like resins. In a blinkered fashion chemists had established the wrong hypothesis; they did not expect a sticky resin, they hoped for a pure crystalline substance.

Baekeland himself was also looking for something else at that time, a substitute for shellac varnish, but he was the only one to perceive the potential in the gluey deposit. In 1907 he discovered that if a catalyst was added to the thermoplastic phenolic mixture and then heated to a much higher temperature, the material produced was so hard that there was no way to soften it. In other words, it had become a cross-linked thermosetting resin.

The first patent describing the method for making this resin was filed in February 1907, and in July a second patent was filed giving the name of 'Bakelite' to this new phenol-formaldehyde resin. These two patents for phenolic resin were the first out of an eventual total of 119. Baekeland's patents expired in 1927 and the market opened up with the registering of a great number of new phenolic trade-names. Since the fifties production of phenolic resins has declined.

In 1910 Baekeland formed the American General Bakelite Company and went into commercial production using a new pressure-moulding process. In 1922 the Bakelite Corporation was launched, and in 1939 it became part of the Union Carbide and Carbon Corporation.

In 1926 Baekeland set up the Bakelite Corporation of Great Britain in England, soon to become Bakelite Limited. Bakelite Limited merged with Spill and Merriam's celluloid company, the British Xylonite Company, to form Bakelite Xylonite, now owned by BP.

Important discoveries are not usually isolated events, and many minds are often trying to solve similar problems at the same time. In England James Swinburne had reached the same conclusions as Baekeland but on going to the Patent Office to file his own patent he discovered that Baekeland's patent predated his by one day. Nevertheless, in 1910 he began to manufacture phenolic lacquer in his own factory in Birmingham, the Damard Lacquer Company which was bought out by Bakelite Ltd in 1928. Damar, or dammar, is the resin from Malay and Indonesian pine trees used to make enamels and lacquers.

Both Baekeland in America and Swinburne in England initially produced phenolic as a coating material for the varnish and lacquer industries, particularly for the protection of metals such as brass, nickel, silver and copper. Later, it was taken up as an adhesive for bonding plywood and other materials, and, finally, as a moulding powder.

In the twenties the electrical, radio and car industries were growing fast and 'Bakelite' appeared to be an ideally tough, dielectric plastics material for moulding all sorts of components. The first telephone part to be moulded in phenolic was a replacement for the bell-shaped hard rubber earpiece attached to the hardwood box of Western Electric's receiver. The age of synthetics had truly arrived when the celluloid billiard ball was replaced by a 'Bakelite' one, and when the white keys on the piano, once ivory, then celluloid, became 'Bakelite' too. Composition golf club heads made from 'Textolite' phenolic by the General Electric Company sent the balls travelling much further than the old wooden clubs. Thomas Edison adopted Union Carbide Corporation's 'Bakelite' for his phonograph records, replacing the old pressings of wax and shellac, although shellac did continue in use until the late forties.

In his book *Plastics History—USA* J.H. DuBois described Baekeland's most difficult challenge as the education of the plastics moulding companies in the correct use of the new thermosetting material, familiar as they were with early thermoplastics. Throughout the years that plastics have been considered substitute materials they have been subjected to existing manufacturing processes and frequently not regarded as materials in their own right. DuBois describes how in March 1938 the 'Bakelite Travelcade' journeyed around America spreading the word. 'What is this Plastics Business?' was its theme, and it provided lectures, exhibitions and even a sound film entitled *The Fourth Kingdom,* referring to the man-made world—the other three being animal, vegetable and mineral.

Phenol-formaldehyde mouldings, castings and laminates

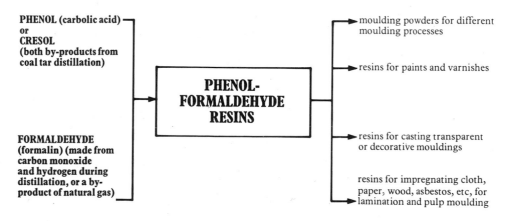

Phenolic resins are the result of complicated chemical reactions, and various kinds of resins are made. The basic phenolic reaction is the condensation of an excess of phenol with formaldehyde, promoted by an acid catalyst. The resulting solid resin is a linear and soluble thermoplastic called a Novolak. If this is ground to a powder, and fillers and dyes are added as well as a curing agent, when compression-moulded this resin becomes a cross-linked thermoset. It is called a two-stage resin because of the addition of the curing agent.

Another type involves the use of an alkali catalyst such as ammonia. This starts a three-stage reaction which produces an infusible thermosetting resin, although in the first stage it is thermoplastic. At that stage the phenol and formaldehyde condense into a soluble short-chain resin syrup called a 'resol', used for cast phenolic laminates and adhesives. In the second stage, when more heat is applied, the polymer chains grow longer. The resin is now called a 'resitol' and is made into moulding powder. The final stage occurs when the resin is taken to an even higher temperature in the mould, which cross-links it into an irreversible thermosetting moulding. The resin at this last stage is called a 'resit'.

Baekeland's original 'Bakelite' resin patent referred to the uncured thermoset resin, the Stage A or thermoplastic Novolak type.

Phenolic mouldings are easy to recognize, being primarily dark in colour, with the reinforcing filler material usually visible as a mottled surface. The main advantages of phenol-formaldehyde are its good electrical and moisture resistance and its general stability. It can be very brittle, and mouldings can fade dramatically due to its unstable pigmentation, leading to its restriction to the familiar dark colours: black, brown, maroon, green and blue. The brittleness of the resin is overcome by the use of the various fillers which can be used decoratively, like simulated walnut or the mahogany graining on radio and TV cabinets. Different fillers impart various properties to the resin. Wood flour is the most common filler for general electrical and heat-resistant mouldings: plugs, sockets and electrical fittings, radio and TV cabinets, saucepan handles, tobacco and cigarette boxes, ash trays by the million, fuse and meter housings, and countless other uses. Other fillers used are cotton flock, asbestos, mica, glass fibres and soya bean (in the forties), all of which improve the mechanical and dielectric properties.

The usual processes for moulding phenolic resin are compression-, injection- and transfer-moulding. The resin in powder or pellet form is already partly cured. The final curing takes place quite quickly in steel moulds under pressure, although the thicker the section the longer the curing time. The moulding is ejected while still hot and any excess flash removed later.

Highly efficient screw threads are easily formed in phenolic mouldings, and many such screw-top containers of the thirties are superior to modern mouldings. Snap-closing 'Bakelite' boxes made their appearance at the same time.

Advertising and Promotion Premiums

The Mennen Company in America was the first firm to introduce screw-on black phenolic tops to their soft tubes of creams in the early thirties. The tops provided an important vehicle for manufacturers to advertise their names and trade-marks. That infectious phenomenon, the premium, was born around 1933 following the Depression: free give-aways are more acceptable in times of poverty. Premiums reflected improvements in mass-production techniques. Today, such objects tend to be so cheaply manufactured that they quickly end up in the dustbin and simply contribute to the debased opinion of plastics people hold. However, at that time manufacturers were concerned that the usefulness of their premiums should be a good advertisement for the product. In their own way, too, these premiums—measuring spoons, beakers and pens—were commercials for the new materials coming onto the market, especially for the favoured materials of phenol-formaldehyde and urea-formaldehyde. As plastics became available at a more reasonable cost, so more and more of these premiums were churned out. A National Premiums Exhibition held in Chicago in 1935 emphasized their importance in increasing sales and introducing new materials.

Decorative Effects in Phenolic

In addition to the moulding of names and trade-marks into the product itself, patterns could also be sandblasted afterwards onto the article to create a contrast

opposite
Examples of phenolic mouldings, including: TV12 table receiver, 1949, Bush Radio Ltd, moulded by British Moulded Plastics Ltd; Smith 'Sectric' electric clock; tasselled celluloid lampshade on 'Goltone' phenolic base; electric bed-warmer, 1947, R. A. Rothermel Ltd; 'The Wunup' spring-loaded cigarette case; classical-style revolving 'Smoker's Friend', mid 1930s; spring-loaded 'Magic Pin Box'; cigar box with dark red lid moulded with a relief of nymph and satyr, 'UDA Bristol'; also mottled bowl, ash try with cast phenolic cigarette holder, tea caddy, light switch and socket. (Author's collection. Photo Fritz Curzon)

Ekco SH25 Superhet Wireless moulded in 1932 by E. K. Cole Ltd, with phenolic panels. (Council of Industrial Design, London)

'Who Said Depression?', 1935. (*Modern Plastics*)

between the smooth shiny plastic surface and the textured design. Such designs are often found on 'Bakelite' boxes.

Another decorative effect, though very rarely seen nowadays, is the use of metal inlays. One mould could produce boxes of very differing appearance not only by using different coloured resins, but by placing various metal inlays into the mould. Enamelled Art Deco shapes or filigrees of copper, gold and silver were laid into the mould before adding the powder, a very popular American technique in the thirties. Plastics were also inlaid after moulding, as in the doors of the Kansas City Municipal Auditorium by Oliver Bernard or in his cocktail bar of the Strand Palace Hotel, London.

Despite the popularity of the later pastel shades of urea-formaldehyde, introduced in 1929, the fashion was for black mouldings, quite alien to today's preference for bright colour. By the mid-thirties three quarters of all mouldings were black. More recently cosmetics firms have readopted a black house-style, especially the Biba range of jars, mirrors and powder compacts.

Wireless Sets

Because of its good insulation, moisture-resistance and shock-absorbing properties, phenolic resin was ideal for the replacement of expensive timber cabinets for radio and similar equipment. Radios in particular suffered shocks through packaging and handling and were affected by changes of temperature.

The first radio cabinets with highly decorated Baroque-style mouldings and fretwork panels appeared in the late twenties. The EKCO SH 25 Superhet Wireless of 1932 is assembled from simpler monolithic textured panels, with the main emphasis on a hand-fretted circular design of trees against a winter sky with watery reflections below. Nearly 40cm. wide (16in.), it was the largest plastic cabinet of the time. It was also the first radio to print both station names and wavelengths on its circular tuner. The tapering form of the whole cabinet eased ejection from the mould.

E. K. Cole Limited (trade-mark 'EKCO'), of Southend-on-Sea, Essex, pioneered plastics cabinet design in England by commissioning work by industrial designers, starting in 1933 with Serge Chermayeff, whose 'Bakelite' designs still betray the

influence of timber. In 1934 they employed Wells Coates, who designed the first circular cabinet; in 1935 Chermayeff and Coates worked together, and in 1936 Jesse Collins of Industrial Designs Partnership brought a new simplicity to cabinet design. In 1937 Misha Black, also from Industrial Designs Partnership, used the latest 'spinning wheel' tuning to produce a cabinet with clean, sweeping lines. Their designs stood out amongst the mainstream of Gothic cabinets in imitation wood and other shapes and forms still rooted in the past.

E. K. Cole Limited sold over 100,000 sets of Coates' 'AD65' radio cabinet in 1935, during what they described as the 'radio season'. Standing 14 inches high it was one of the largest mass-produced mouldings on the British market. This was bettered in America by the new All-Wave Pilot Radio designed by Jan Streng, which was over 18 inches high. This somewhat Gothic-style radio cabinet was manufactured in phenolic by the Makalot Corporation, for the pioneering radio cabinets had proved that the initial expense for tooling-up for production was an investment.

Two walnut-finish phenolic radio cabinets designed by Wells Coates; *right* the first circular cabinet, 1935, 35.5cm (14in.) in diameter, *left* 35.5 × 45.7cm (14 × 18in.), both for E. K. Cole Ltd. (*Modern Plastics*)

The New Pilot All-Wave Radio designed by Jan Streng, 1936, 34.2 × 26.3 × 46.3cm (13½ × 10⅜ × 18½in.), moulded in 'Bakelite' and 'Beetle' resins by Associated Attleboro Manufacturers Inc. for the Pilot Radio Corp., USA. A single compression-moulding in a 500-ton press. (*Modern Plastics*)

Furniture

The earliest record of plastics in furniture is a Regency-style armchair manufactured in 1926 by the Simmons Company, of Racine, Wisconsin. It was constructed from eight separate phenolic mouldings reinforced with green fabric, with an upholstered seat and back on a timber frame. Identical designs moulded in the latest plastics are still fast sellers.

In contrast, French phenolic furniture designs in 1937-8 seem worlds apart from reproduction mouldings in America. Tubular cantilevered chairs displayed a post-Bauhaus functionalism as hygenic school furniture. Desks combined with chairs to form rationalized and uncluttered units with clean simple lines, refreshing to the eye after the imitation Baroque and Regency popular then in the States.

Phenolic furniture moulded by Manufacture d'Isolants et Objets Moulés was first made in France in 1932 for outdoor cafés and terraces. It enjoyed great commercial success. The tabletops, seats, backs and armrests were moulded in phenol-formaldehyde in plain and mottled colours, primarily black and the darker shades of red, brown and green. The underside of the table had reinforcing ribs moulded in to give rigidity, and metal inserts into which the legs were screwed. Such thick heavy sections, needing 1,500-ton presses, would be quite uneconomical today.

Occasional chair developed by Ed O. Wokeck, 1926, for the Simmons Co., USA. Eight phenolic mouldings reinforced with fabric; seat and back on timber frames. (Courtesy Ed O. Wokeck)

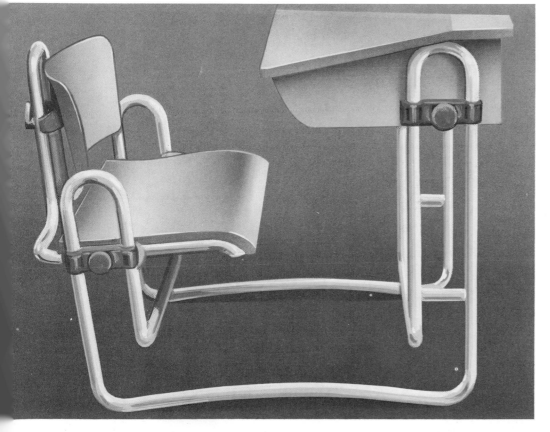

above
French café furniture, 1932, moulded in solid phenolic by Manufacture d'Isolants et Objets Moulés. The tabletop required a 1500-ton press. (*Modern Plastics*)

School furniture exhibited at the Salon des Arts Ménagers, Paris, 1937. Only a prototype, but remarkably modern in its functional approach, with chair back and desk top in plastic. Designed by Jacques André, manufactured by Jean Prouvé. (*Modern Plastics*)

Phenol-formaldehyde Laminates

Laminates for industrial uses are usually made of phenol-formaldehyde, whereas decorative laminates have surfaces made from lighter-coloured urea- and melamine-formaldehyde resins. Laminated phenolic has much greater impact strength than moulded phenolic and in the past was used for heavy-duty applications such as gears, wheels, pulleys and propellers. Early experiments were carried out with the impregnation of long fibres of jute and sisal. Henry Ford's first plastic car body of 1941 was made from pressed panels of laminated soya bean, 30 per cent phenolic resin and 70 per cent fibre, making very light-weight and economical mouldings.

Impregnated cotton cloth was laminated to build propellors and gears as it could be bent and stretched round curved shapes. Great skill was necessary in designing the tools, as care had to be taken to choose the correct directions for the laminae to lie for maximum strength. Phenolic laminates were pioneered by Dr Redman at Westinghouse in 1910 and enjoyed considerable use during World War I. Heavy canvas impregnated with 'Bakelite' was moulded into gears for the internal combustion engine and found to be self-lubricating and amazingly noise-free, as well as shock-absorbant and unaffected by oils and acids. The replacement of metal by plastics is considered to be a relatively recent phenomenon, for example in cars, but it started over sixty years ago.

Before the arrival of urea resins, the Formica Insulation Company in America founded by Dan J. O'Connor as a Westinghouse subsidiary, developed decorative laminates for radio cabinets using dark phenolic laminates bonded to a top layer printed with different simulated wood grains.

By the late twenties phenolic laminates had reached the domestic market despite their dark colours. Later to be replaced by the lighter coloured ureas when they appeared, they were used for draining-boards and all kinds of decorative panelling. The Regent Palace Hotel, London, possessed a fine circular cocktail lounge, designed by Oliver P. Bernard, now alas dismantled, designed in 1935 in the 'American style'. Everything possible was made of gleaming Formica, from tabletops and wall surfaces to the armrests of the couches.

World War II brought a tremendous impetus to the use of phenol-formaldehyde laminates, particularly in aircraft: aerial masts, ducting, brackets and pilot's handwheels. The seats of Spitfires were made from low-pressure moulded phenolic impregnated fibres, as well as from fabric, timber veneer and even manilla paper. Parts of boats previously made of the hardest available timber, lignum vitae, were replaced by laminated phenol-formaldehyde. In 1949 the General Electric Company was laminating 'Texolite' kickplates. And later, in 1956, the first plastic re-entry nose-cone laminated from glass- and asbestos-filled phenolic was fitted to a Vanguard missile.

Phenolic laminates in sheet, rod or tube were extremely dense, with polished or matt surfaces in shades of black and brown. Sheets were laid up by hand or machine, then passed through a resin bath and pressed between rollers. The solvent in the resin (industrial alcohol) was evaporated and layers of impregnated sheets were pressed between the hot polished plates of a hydraulic press, and finally heat-cured.

Tubes and rods were made by rolling the impregnated material around a mandrel before curing in an oven, after which the mandrel was removed or collapsed. Tubes formed round a mandrel could also be cured under pressure and heat inside a mould.

Phenolic laminates could be machined on standard timber machinery and an excellent finish achieved. Gears, cams and wheels of all kinds were fabricated from sheet, while other parts were cut from tube with screw threads machined into them.

Cast Phenolic

Most decorative designs were made by the third method of processing phenolic resin, casting, a process rarely used nowadays. Unlike the resins used for moulding and laminating which are reinforced with fillers, making them opaque, cast resin could be made more attractive because no filler was used.

Evening wrap designed by Helen Runyon of the Revolite Corporation. It won first prize in the Style Group, First Modern Plastics Competition in 1936. Quilted silver fabric treated with 'Bakelite' phenolic resin, washable and non-tarnishing, it is the first recorded use of an industrial resin for evening wear. (*Modern Plastics*)

Electric fan by the Westinghouse Electrical and Manufacturing Co. Ltd, 1937, with propellors laminated from 'Micarta' phenolic-impregnated fabric. Much quieter than metal blades, non-corrosive and non-warping. (Collection Dr and Mrs Nicholas Kemp. Photo Fritz Curzon)

Casting resin was made with 'resol' or 'resitol' syrups containing a higher proportion of formaldehyde, treated to retain its pale yellow translucent character.

The phenolic resin syrup was poured into heated, open lead moulds; in the forties a flexible mould of elastomer gave a more polished finish to the surface. A catalyst was added to speed hardening, along with whatever pigments were required. The polymerization reaction could take up to a week to complete, with the moulds being maintained at a temperature of 150°-175°F. Mixtures of two or more colours produced the well-known gem-like effects of rose quartz, onyx, jade, amber, ruby, turquoise and tortoiseshell. Black cast phenolic simulated jet. Usually these resins were opaque or translucent, but the addition of a modifier could produce clear castings. Bakelite's transparent phenolic grade XM 9131 could be boiled and sterilized and therefore found many medical and dairy uses.

Familiar cast phenolic objects of the thirties include toilet articles and jewelry, hairbrush sets, earrings and bracelets, buttons, and umbrella handles.

Cast tubes and rods were transformed into jewelry, often making a co-ordinated set in one colour. Stock sections in the form of polished or unpolished sheet, tubes, rods and blocks of all shapes were sliced up like salami and machined on standard metal- and wood-working equipment. The pieces were shaped like sculpture using small abrasive wheels, and cutters, and then buffed and polished.

More recently these decorative uses of cast phenolic have been superseded by cast acrylic, producing, for example, the same swirling onyx or alabaster patterns on door-knobs, cupboard handles and finger plates.

Cast resins were sometimes put to unexpected uses such as armrests on chromium-plated tubular chairs, or the cast rods often found on candlesticks and chandeliers. Wood coated with transparent liquid 'Catalin' produced an effect known as *bois glacé*, which was very popular for jewelry and also replaced the glass tops on desks.

Streamlining in plastics is an inherent characteristic of the material. Plastic objects are curved because the polymers need to flow within the moulds and corners are difficult to produce. In addition, compound curves and domed shapes are fundamentally stronger than flat areas, as they are better able to distribute the stresses. Bevis Hillier pointed out in his study *Art Deco* (Studio Vista, London; Dutton, New York 1968) that Aztec art was one of the many influences on design in the twenties and thirties, and we do associate ziggurat shapes and geometric forms with that monumental and modernistic period in architecture and design. However, in plastics these stepped shapes with rounded fluted details are as much a result of plastics technology as they are of design influences. They are the products of the jelly-mould principle: to remove an object from a mould there needs to be a certain amount of taper, with at least some slight curving of the corners, while the more ribbed the surface, the greater the strength imparted to flat areas and thin sections.

Concealed function and misleading appearances were also a predominant feature of mouldings in the twenties and thirties. Admittedly our aesthetic appreciation has changed during the last forty years, yet it is nonetheless disconcerting to find pepper in what looks like a sugar dispenser. What looks to us like the streamlined rear lamp of a motor car is in fact a watch display case. But there are hazards if the streamlining tendency is taken to extremes, as Alan Jarvis pointed out in *The Things We See* (Penguin Books, Harmondsworth 1946). Objects were often fatally placed on the curved edges of stream-lined fridges.

UK: Bakelite (Bakelite Xylonite), Carvacraft and Gaydon (John Dickinson and Co.), Catalin (Catalin Ltd), Delaron (De La Rue Plastics), Erinoplast (Erinoid Ltd), Mouldrite PF (ICI), Nestorite (James Ferguson & Sons), Paxolin (Micanite and Insulators Co.), Rockite (F. A. Hughes and Co.), Sternite (Sterling Moulding Materials)
USA: Bakelite (Union Carbide Corporation), Catalin (Catalin Corporation), Durez (Durez Plastics and Chemicals), Durite (Durite Plastics), Fiberite (Fiberite Corp.), Heresite (Heresite and Chemical Co.), Indur (Reilly Tar and Chemical Corporation), Makelot (Makelot Corporation), Marblette (Marblette Corporation), Opalon (Monsanto Chemical Co.), Prystal (Catalin Corporation), Redmanol (Union Carbide Corporation), Resinox (Monsanto Chemical Co.), Revolite (Revolite Corporation), Textolite (General Electric Co.)
Trade-names for phenolic laminates include:
UK: Bakelite (Bakelite Xylonite), Formica (Formica Ltd), Paxolin (Micanite and Insulators Co.)
USA: Bakelite (Union Carbide Corporation), Formica (Formica Insulation Co.), Micarta and Westinghouse (Westinghouse Electrical and Manufacting Co.), Textolite (General Electric Co.)

AMINO PLASTICS

The amino plastics are resins made from compounds derived from ammonia. The carbon dioxide and ammonia ingredients in synthetic aminoplasts were originally obtained from coal but now come from natural gas. If the fossil fuels of the world were depleted, the two ingredients could be obtained from air and water.

There are two kinds of aminoplasts, urea-formaldehyde and melamine-formaldehyde, both organic amino compounds; like phenol-formaldehyde, they are both thermosetting plastics resulting from condensation polymerization.

Urea-formaldehyde (UF) *thermosetting*
Urea-formaldehyde, also called urea resin, is the amino plastic made from the

synthetic amino compound urea ($NH_2\ CONH_2$) often used as fertilizer, condensed with formaldehyde. The standard moulding process is compression moulding, and typical mouldings are the old and familiar cream or green picnic sets, early cream or red telephones, modern coloured toilet seats, and white and ivory electrical fittings. Urea-formaldehyde came on the market in Britain in 1924. In contrast to phenolic resin, it was the only plastic material available to make white or light-coloured mouldings cheaply until the arrival of polystyrene and polythene, and so its impact was enormous.

Like phenolics it is a rigid durable thermosetting polymer with good dielectric strength and chemical resistance. It can be produced in translucent colours as well as opaque, thereby giving it a wide market as lampshades and light diffusers. Unlike phenolic it does not discolour with age, but some of the early mouldings have faded, probably due to faulty moulding.

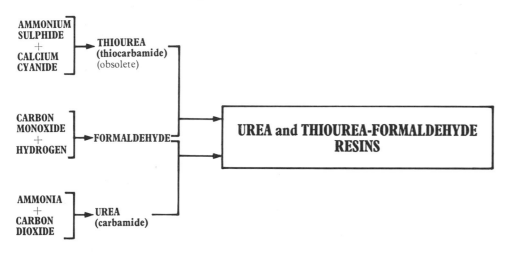

1) UF resin from the first stage in the moulding process is used for-
(a) rigid foam (b) paints and enamels (c) laminated panels from impregnated paper or fabrics
(d) adhesives for bonding plywood and treating fabrics

2) In the second stage of the moulding process, the resin becomes solid and is made into moulding powders for all kinds of mouldings

3) The third stage cures the UF resin under heat and pressure into a thermoset

All the end products are cured; for example, glued plywood cures under heat and pressure.

As the chart shows it can be supplied as a powder for moulding and as a resin or adhesive for laminating panels and impregnating fabric. Non-crease chintz is glazed with urea resin, and urea enamels used to be baked onto cookers, refrigerators and similar equipment. Foamed urea is now very popular as thermal cavity insulation in the building industry. The liquid resin is injected into cavity walls where it expands *in situ* to form a rigid foam.

In the late nineteenth century it was known that urea and formaldehyde would react together and condense to form a transparent resin, the first stage in the production of urea-formaldehyde. In 1915 Dr Fritz Pollack in Vienna, searching for a substitute for glass, took the experiment further and attempted to cast this syrupy resin into a transparent solid. He did not succeed, for his 'glass' crazed and bubbled, but he nonetheless filed various patents from 1920 onwards. This particular line of research was not resolved until acrylic appeared in 1933.

However, the breakthrough came in England in 1924 when the first successful light-coloured moulding powder was made by Edmund Rossiter, chief chemist at the British Cyanides Company, and used by the Beetle Products Company who called the material 'Beetle' thiourea. It was displayed at the Wembley Exhibition in 1925 and hailed as the new white hope. By 1929 thiourea products were well established on the market under such trade-names as 'Bandalasta', 'Beatl', 'Beetleware', 'Birmite' and 'Linga-Longa'. By that time 'Beetle' urea-formaldehyde was also available, and by 1931 thiourea and urea had achieved terrific success across Europe.

Thiourea is very similar to urea, differing only in having one atom of sulphur in its molecule. Rossiter discovered that it was necessary only to add similar fillers and pigments to those used in phenolic to make it into a stable, mouldable plastic. Wood flour is the most common filler used with urea.

The American Cyanamid Company started to produce 'Beetle' UF in America. The Bakelite Corporation manufactured its own urea, and then around 1929 'Plaskon' urea-formaldehyde appeared, made by the Allied Chemical Corporation and 'Uformite' produced by Resinous Products and Chemical Corp. By 1931 there were many other urea trade-names in America, such as 'Aldur' (America's first clear urea resin) and 'Durez'. By the early thirties urea had more or less superseded thiourea-formaldehyde.

It is difficult nowadays to imagine the impact of the new material. In the space of a few years it had filled homes with colourful mouldings. The colours were, however, what Roland Barthes has called 'chemical' colours: pastel shades of blue, green and orange, and British Cyanides even supplied over forty different shades of cream and white. Picnic baskets, cruet sets and vacuum flasks, snap-fastening boxes for cuff-links, talcum powder containers and cases for dry shavers, all sprouted in the new material. In America Mickey Mouse teasets in 'Beetleware' sold by the thousand.

In 1929 telephone design was totally revolutionized by the new synthetic materials. In that year the candlestick telephone, which had been with us since around 1917, with its separate hook-on receiver made of brass coated with ebonite (hard rubber), was replaced in America by the 'French phone' desk set on its circular base, and in Europe by the Siemens 'Neophone' hand micro-telephone, a design still in use today. Millions of 'Neophones' were moulded in black 'Bakelite' phenolic and rather less in urea-formaldehyde in cream, red, green and brown. The brown is very rarely seen. Silver and gold-sprayed handsets were also available as standard models. By 1941 a coloured telephone was still something special, and Diana Dors wrote at age 9, 'I am going to be a film star, with a swimming pool and a cream telephone.'

The main disadvantage of thiourea and urea is their tendency to absorb water. In 1939 an improved formaldehyde, melamine (MF), which was not affected by this problem, was produced commercially. Thiourea was totally abandoned, and melamine-formaldehyde plastics became widely used for decorative laminates and moisture-resistant moulding materials. Where urea did come into contact with liquid, for example, as picnic cups, eggflip mixers, ribbed draining boards and hospital trays, it is often found to have crazed quite badly.

The beautifully mottled and multi-coloured effect of thiourea and early urea mouldings is a treatment rarely seen in plastics today except in jewelry and cosmetic containers moulded in the newer plastics.

Case for the Toledo weighing scales, 1934, winner of the Industrial Group, First Modern Plastics Competition, 1936. Moulded in 'Plaskon' urea, replacing the previous Toledo balance made of porcelain. Developed by Harry E. Hire and James L. Rogers for the Toledo Scale Co. (*Modern Plastics*)

In 1936 the American journal *Modern Plastics* sponsored the first Modern Plastics Competition. Winner of the industrial group—and hence recipient of a urea moulded plaque—was the 'Toledo' weighing scales, the most ambitious, dramatic and influential moulding to emerge at that time. Its influence on machine housings and cabinets can be seen in all designs since; it was the grandfather of modern product casings, right down to structural foam electric typewriter housings. Designed in 1934, it was the outcome of five years development on a grandiose scale. Formerly shop scales and weighing machines were very decorative cast-iron balances, lacquered with gold foliage and lettering and with separate weights. 'Plaskon' UF moulding material was chosen since its smart white finish seemed most appropriate for a modern hygienic food-weighing machine. But size was a problem for the moulders. After five years research the Toledo casing turned out to be the largest white moulding ever produced in a single operation. In addition, it was moulded on the biggest press in the industry, the first of the giant presses, so like the ominous machine-god Moloch in Fritz Lang's 1926 film *Metropolis*. It had a 36-inch ram and a 36-inch stroke, was nearly two stories high, exerted a pressure of 1,500 tons and weighed 89,000lbs. The mould itself was of a specially hardened steel. Nevertheless, the press produced a moulding that cost 85 per cent less than the previous model.

The largest press of its time for moulding the Toledo weighing machine cover, built by the French Oil Machinery Co. The mould was made by the General Electric Co., 1934. (Photo by courtesy of J. Harry Dubois; drawing from *Modern Plastics*)

Probably the earliest children's high chair to be made of plastics was designed by A.H. Woodfull, manager of the product design department at BIP until his recent retirement. It won first prize in the 1951 Horner's Award, and was conceived before the shape of plastic chair-shells became familiar, such as Robin Day's polypropylene chair of 1963, and even before the Eames GRP chair of 1949. The chair was body-shaped, and fitted to a tubular steel frame. An anti-germ curved tray was moulded in melamine to resist heat, stains and scratches, and was removable, unlike any tray at that time. The arms could then be pushed back inside the tubular frame. The particularly advanced form of this high chair brought new materials into what was traditionally a market for solid timber, producing a children's chair onto which 'the hose-pipe can be turned'.

Hi-Gene High Chair designed by A. H. Woodfull, which won the 1951 Worshipful company of Horners Award. This original hand-made prototype was made from urea and melamine materials produced by BIP Ltd.

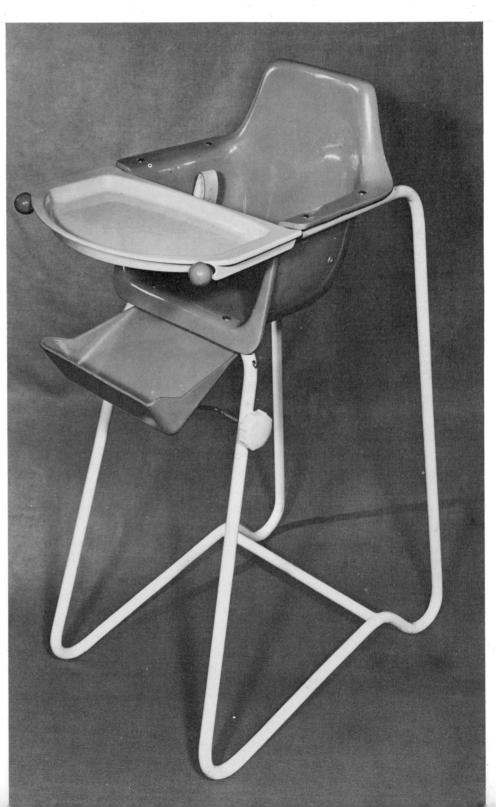

Urea-formaldehyde Decorative Laminates

Industrial laminates were made up from phenolic impregnated sheet. The other type of laminate, with which most people are more familiar, is the coloured decorative laminate such as 'Formica', made with a top surface of melamine-formaldehyde. Until melamine took over in 1938, early decorative laminates were made from paper or cloth, impregnated with urea or thiourea resin. With the solvent removed, the impregnated sheets were flattened and cured under heat in a hydraulic press. Different grades of papers and fabrics were used, and the final surface texture was applied either by embossed plates on the press or integrated in the stack with patterned paper as the top layer. A similar process, for example, incorporating grained timber or marbled effects, is still used for producing 'Formica' melamine laminates today. Urea laminates could be cold-bent if reasonably thin, or heat-formed into curves and cylinders, and machined.

The hygienic plastics-surfaced kitchen of today was virtually unknown until after World War II. In 1949 at the Electric Development Exhibition a kitchen was displayed with curved cupboards and drawers faced with laminated urea-formaldehyde. The moulded sink, draining-board and china cupboard were in acrylic ('Perspex'). This was 'no idealistic architect's dream, but a composite view of an actual kitchen erected at an electricity exhibition' (H.R. Fleck, *The Story of Plastics*, London, n.d.).

Urea laminates brought functional simplicity to a multitude of designs, from wall panelling, cupboard fronts and dinette tables to the bars and restaurants of the most exclusive liners of the thirties.

Bonding Plywood

One of the most spectacular successes of urea was its use as a bonding resin for plywood during World War II. Urea-formaldehyde resin helped evolve new construction techniques, increased the rate of mass-production and made aircraft production cheaper and more efficient.

During World War I the standard glue for the early all-wood aeroplanes was casein, or animal and albumin glues. A major disadvantage was that flying in very

All plastics kitchen with units constructed from urea-formaldehyde laminates and 'Perspex'. The Electrical Development Exhibition, London, 1949. (British Electrical Development Association)

cold temperatures made the glue so brittle that the whole plane could collapse under stress. In addition, these glues eventually became rotten.

The development of urea-formaldehyde cold-moulding synthetic glues around 1931 was a closely guarded military secret, and remained so until the end of the war. Together with a new electrical bonding method the advent of this glue made possible the design of such planes as the De Havilland 'Mosquito' fighter-bomber, the fastest war bomber, which first flew in 1940 and which was constructed entirely out of plastic-bonded plywood. It was subjected to air speeds of over 400 m.p.h. at high altitudes and in both arctic and tropical conditions. Urea resin was also used in the construction of the Horsa Glider, as well as for boats, rafts and pontoons.

Using a technique in which a rubber sheet is vacuum-sucked over sheets of hardwood ply veneers coated with resin and laid over a mould, it was possible to shape much larger mouldings than ever before. Complete sections of aircraft, wings and fuselages were laminated in this way.

The new electrical high-frequency welding process swept aside the lengthy moulding cycle of steam, heat and pressure: it cut the time for bonding wood veneers from between ten and twenty minutes to a mere three minutes. Aircraft tailplanes, previously constructed in jigs and held together by thousands of rivets, were now laminated in a fraction of the time with some thirty rivets integral with the moulding.

The technique was not only useful for bonding wood to wood, but successfully bonded wood to metal, rubber, fabrics, plastics and even glass.

In 1940 a competition was held in America, sponsored by the Museum of Modern Art, New York, entitled Organic Design in Home Furnishings, of which the joint winners were Charles Eames and Eero Saarinen. Their joint design was executed in plywood formed with 'Zenaloy', a new synthetic resin, and was also an entirely new approach to furniture construction. Up until then wood had only been bonded in two dimensions, like the steam-and-pressure-formed furniture of Michael Thonet in the 1850s and Marcel Breuer and Alvar Aalto in the 1930s. High-frequency welding, the rubber bag process and synthetic glues gave plywood plastic qualities enabling it to bend and stretch in several directions at once and finally to set in compound curves. The process is somewhat similar to drape-moulding thermoplastic sheet. As with phenolic laminated fabric or decorative laminates, the fine ply veneers gain strength when impregnated and cured with a thermosetting resin.

Nowadays urea adhesives are still widely used in the furniture industry and nearly all particle board and blockboard is bonded with urea resin.

UK: Aerolite (Ciba-Geigy), Bakelite Urea (Bakelite Ltd), Beetle (Beetle Products Co., now British Industrial Plastics. Beetle, Bandalasta, Beetleware, ML and Linga-Longa were all products made in 'Beetle' resins), Mouldrite U (ICI), Nestor and Nestorite (James Ferguson and Sons), Pollopas (British Industrial Plastics), Scarab (Beetle Products Co.), Traffolyte (De La Rue), U Foam (ICI)
USA: Bakelite Urea (Union Carbide Corporation), Beetle (American Cyanamide Co.), Cascamite (adhesive, Casein Co.), Durez urea (Durez Plastics and Chemicals), Plaskon (Allied Chemical Corporation), Uformite (Resinous Products and Chemical Corp.), Uralite (Consolidated Moulded Products)
GERMANY: Kaurit adhesive
AUSTRIA: Pollopas and Aminolac (Etablissements Kuhlmann)
SWEDEN: Skanopal (Perstorp AB)

Melamine-Formaldehyde (MF) *thermosetting*
Melamine-formaldehyde is very like urea-formaldehyde but as the structure of its cross-linking is much more intricate the polymer has a much greater water resistance, is much tougher and more scratch-resistant, and possesses improved heat and chemical properties, all of which indicate its suitability for tableware of all kinds, and the handles and knobs of electrical appliances. It is capable of being brightly pigmented and given a high gloss finish. Although urea is a good general moulding material, melamine is better in all respects, except that it is more expensive.

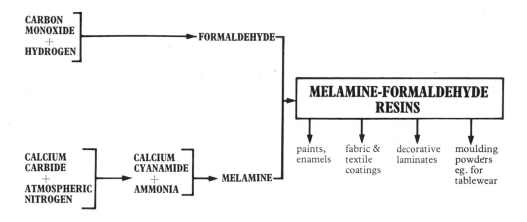

Melamine resin was first discovered in Switzerland by Baron Justus von Liebig in 1834, but it was not until a century later that it was taken up again. The first patent for reacting melamine with formaldehyde was filed in 1935 by Henkel, although it was 1939 before the American Cyanamid Company brought melamine moulding resins onto the market. Like phenolic resin, melamine resin was first marketed in the early thirties as a synthetic resin for glueing, coating and enamelling but at first was only used for the priority purposes of war and was not generally available. Even by 1947 its main use was still in the manufacture of tough hard-gloss paints and enamels, and crease-resistant fabrics.

As the heat and abrasion resistance of melamine is greater than that of both phenolic and urea resins, it is ideally suited for the making of decorative laminates, with the familiar household names such as 'Formica', 'Arborite' and 'Warerite'. Because it is more expensive than phenolic only the top sheet of the lamination bearing the pattern is impregnated with melamine resin. It is placed on top of phenolic-impregnated layers and the whole block is hot-pressed.

Melamine laminates gradually took over all the roles previously filled by phenolic and urea for decorative surfaces, such as counter-tops and restaurant tables. 'Micarta' laminates were developed in 1937-8 by Westinghouse as 'An amazing new plastic surfacing material—beauty, utility, durability'.

Warerite Limited, of Ware, Hertfordshire, laminated genuine tapestry into melamine panels for decorating the interiors of liners and railway carriages. 'Formica' was first manufactured by the Formica Insulation Company of Cincinnati, Ohio, in 1913 but only became popular in the UK in the late fifties when it helped to create new American-style 'luxury' kitchens.

Most mouldings of melamine-formaldehyde are compression-moulded with a filler of paper pulp. The first mouldings in melamine were produced during the war for US Service buttons and tableware, and melamine has not only helped spread the domestic acceptance of plastics in the home, but has also stimulated new approaches to tableware design. A pinic set for four people, created by Helen von Boch and Federigo Fabbrini for Villeroy and Boch, Mettlach, Germany, stacks ingeniously together to form one portable unit. Other sets have been designed to stack together into a variety of shapes on the same principle. Melamine dishes weigh some 70 per cent less than the same quantity of pottery, and are machine-washable.

Many people dislike melamine tableware, possibly because it has associations with children's cups and plates which are often embellished with nursery transfers and Walt Disney favourites. These transfers are applied by placing the printed foil down on the partly cured moulding, and closing the mould again to complete the fusing and curing. Tough as it is, moulded melamine can crack if thrown onto a hard surface, unlike polycarbonate baby bottles, for example, which simply bounce back.

The warm, glossy, high-quality finish on melamine is often the main reason for choosing it for a specific application, despite its increased cost. The heavier and more

Bomba picnic set, 1973, designed by Helen von Boch and Federigo Fabbrini, for Villeroy and Boch, Mettlach, Germany. 39.3 × 25.4cm (15½ × 10in.) when packed away, as in upper right of illustration. (Villeroy and Boch)

opposite
A range of mouldings dating from the 1880s to the present day in phenol-formaldehyde, vulcanite, urea-formaldehyde, cellulose acetate and polythene. (Author's collection. Photo Ken Randall)

substantial a plastic object is, the more positive appeal it seems to arouse, as if the very lightness of a moulding implies that it is cheap and worthless.

UK: Melaware (Ranton and Co.), Melmex (BIP), Melolam and Melopas (Ciba-Geigy), Beetle melamine (BIP)
USA: Catalin Melamine (Catalin Corporation), Melmac (American Cyanamide Co.), Plaskon Melamine (Allied Chemical Corporation)
ITALY: Melbrite (Montedison)
DENMARK: Mepal (Rosti)
Trade-names for melamine decorative laminates include:
UK: Arborite (Arborite Ltd), Formica (Formica Ltd), Warerite (Bakelite Xylonite)
SWEDEN: Coronet (AB Tilafabriken), Isomin (Perstorp AB)
USA: Formica (Formica Insulation Co.), Micarta Westinghouse Electrical and Manufacturing Co.)

POLYVINYL CHLORIDE (PVC) *thermoplastic*

The term vinyl is used as an abbreviation for many polyvinyl materials, but usually refers to PVC (polyvinyl chloride, $CH_2 = CHCl$). Polymers made from vinyl chloride are compounded with a wide variety of additives to produce a stable range of PVC plastics, varying from very rigid to very soft and flexible. With its ability to copolymerize so diversely, PVC is a versatile material.

The vinyl plastics are synthesized from acetylene, which itself is obtained from calcium carbide, produced from coke and from ethylene, which is derived from oil or natural gas.

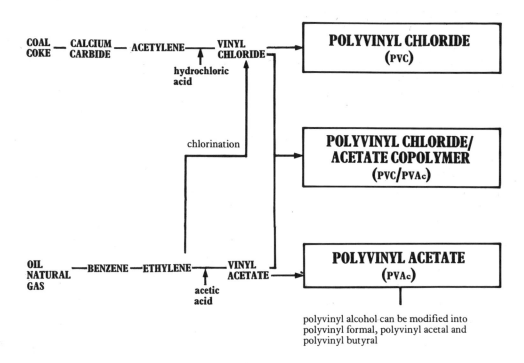

PVC was first documented in 1835 when Henri Victor Regnault observed the formation of a white powder in a test tube. In 1872 E. Baumann first polymerized a hard PVC but it was not until I. Ostromislensky's work in Moscow in 1912 that vinyl chloride was first produced in a soft plasticized form, as a new rubber-like compound. After another long time-lapse PVC technology was revived in Germany, and by the outbreak of World War II there was an established PVC industry, mainly due to the work of Reppe. Encouraged by the growing shortage of natural rubber and by increased demand on account of Germany's preparations for war, Reppe discovered a method of producing PVC on a commercial scale.

America's first PVC was manufactured in 1928 by the Carbide and Carbon Chemical Corporation as a rigid plastic under the trade-name 'Vinylite'. RCA-Victor used it for a while, with the addition of a filler and black pigment, to press records. It was used alongside the old favourite, shellac, and, although more expensive, its sound reproduction was of superior quality. Records are now pressed from polyvinyl chloride/acetate copolymer—PVC/PVAc. Being a thermoplastic, vinyl can be moulded using simple techniques, and flash and scrap can be recycled if clean.

'Vinylite' was also used for impregnating fabric and for casting dentures, and was moulded into steering wheels, toothbrushes or into mottled sheets for making cosmetic jars, cigarette boxes and clock cases. As it was practically tasteless it could be used for moulding tumblers and food containers.

An opalescent effect resulted from incorporating aluminium or bronze dust with

74

the resin. Even 'fish essence' was used to create a shimmering nacreous effect. It must have smelled very unpleasant if accidentally burnt! 'Vinylite' was also used around 1936 for erosion casting in medicine where it replaced celluloid and acetone for making teaching models from human tissue. (See the section on polyesters, p. 130, for erosion casting.)

Modern PVC is available in two grades: unplasticized (rigid) and plasticized (flexible). The former, UPVC, is hard and strong like 'Vinylite'. It weathers well and resists chemical attack. Accordingly it is mainly applied to cold water piping systems and all kinds of drainage and underground fittings where it has replaced cast iron and salt-glazed earthenware. It is also extruded into wall-cladding profiles, all-plastics window-frames and profiled drawer sections. ICI's 'Darvic' sheeting is thermo-formed into clear corrugated roofing. The first flexible vinyl plastic in America was introduced by the B. F. Goodrich Company in 1931. In 1938 other large companies began to produce PVC: E. I. Du Pont de Nemours and Company, Monsanto Company and in 1940 the Dow Chemical Company. PVC was first produced commercially in Britain in 1942.

Plasticized PVC is manufactured in varying degrees of flexibility from soft and rubbery to hard according to the amount of plasticizer in the resin. During World War II it was used for electrical insulation and vehicle tyres, and generally replaced rubber in aircraft, radio and electrical goods. After the war the material was adopted for the domestic market. But the PVC of the thirties and forties was often badly produced: many articles simply failed chemically, becoming extremely brittle with wear or smelling unpleasant. Even now PVC shower curtains still tend to become stiff.

An endless variety of domestic goods use injection-moulded, extruded, blow-moulded, or thermoformed PVC: dolls, balls, shampoo sachets, bottles of all shapes and sizes, shoes, sheeting, trays, leathercloth upholstery. 'Naugaform' PVC fabric is vacuum-formed over Eames' chairs in only two minutes. Vinyl flooring on the market now can be textured to look just like slate, with a raised non-slip surface.

PVC-coated road-signs, fencing and litter bins are maintenance-free, as it is tough and weather-resistant. Tins for canned food, and a wide variety of storage tanks are lined with PVC. Bulk liquid containers made of a nylon-based fabric coated with polymers such as PVC or nitrile/PVC can be used to transport potent liquids like fertilizers, fuels, oils and chemical effluent. Vinyl foils simulating hardwoods, or in plain colours, often replace veneers for timber and chipboard, or cover washable wallpapers and maps. PVC's thermoplastic moulding quality has even been exploited in the making of clothes, and Margaret Fisher has designed a British Rail uniform of knitted jersey in the shape of a tube interwoven with metallic fibres embedded in PVC and polyester. The tube is pulled over a former and shaped in an oven.

In the early fifties extruded PVC tubing began use as a protective sheath on metal furniture. Ernest Race used PVC supplied by BX Plastics Limited on his famous garden chairs designed for the Festival of Britain in 1951, where it was pulled over galvanized springs.

PVC tubing is even used in medicine, though this is now questioned on account of side-effects, since PVC has been found to contain traces of certain compounds which are slowly leaked into the body tissues eventually leading to hepatitis.

In 1974 the vinyl chloride monomer from which PVC is polymerized was declared a health hazard after the death of two workers who had handled the monomer. This hazard only applies to the unpolymerized material and not to the manufactured end-product, fortunately for the makers of PVC food containers. In fact, the most common use for PVC bottles is for non-returnable containers for fruit squash.

Plastisols

Plastisols are PVC pastes for coating fabrics and metals, on which they form a rubbery solid skin. Made of fine PVC powder suspended in various liquid plasticizers, with added pigments and fillers, they are used in various processes such as slush and rotational moulding, dip coating, spraying and film casting.

Boalum lamp designed by Livio Castiglioni and Gianfranco Frattini in 1969, manufactured by Artemide. A snake-like PVC tube with coiled spring reinforcement containing wiring and bulbs, each section 200cm (78½in.) long. A luxury version of a wiring system used to indicate English roadworks at night.

One of the several room-settings constructed for the Pierce Foundation, USA, 1933, illustrating the use of new materials in a three-roomed flat. Everything is plastic, including the transparent 'Vinylite' windows. The walls were made from the largest panels to be fabricated at that time. (*Modern Plastics*)

Multicoloured mouldings in 'Vinylite' resin, 1933, including: mottled sheets behind steering wheel, clock cases, cosmetic jars, record, cigarette boxes with coloured discs in the foreground. (*Modern Plastics*)

The *Elliot Chair* (also called 'Numax'), probably the first pneumatic chair in England, patented in 1942 by Elliot Equipment, specialists in aircraft dinghies. It was manufactured in 'Numax' PVC-coated fabric and could be fitted to a metal or timber frame. (*Architectural Review*)

Most of us are familiar with vinyl-coated metal chairs, draining racks, rubber-backed carpets, bicycle handgrips, or children's wellington boots, all of which are dip-moulded in plastisol. Plaster, cement, wax or synthetic resins such as polyester, can all be cast in flexible plastisol moulds, which are so flexible that they are simply pulled off the cast and re-melted for use. Traffic-controlling 'Glocones' used by the police and road repairers are one-piece dip mouldings made on a former. Leathercloth (PVC sheet on a backing of fabric) is made by spreading plastisol paste on the fabric and heating it. In Sweden PVC is fused onto open-weave cotton and made into scrubbable scratch-proof wall coverings. Plastisols with a blowing agent added make both flexible and rigid PVC foam.

PVC/PVAc Copolymer *thermoplastic*

If vinyl chloride is polymerized with vinyl acetate, vinyl chloride/acetate copolymer results, amalgamating the properties of both compounds in a predetermined ratio. This is a tough, flexible vinyl with good dimensional stability, resistant to impact, chemicals and heat deterioration.

Major uses for this copolymer are in adhesives, flooring and in the pressing of records. As a coating on fabrics, PVC/PVAc has facilitated the development of inflatables and brought a new dimension into everyday living. After the mid-sixties, with the introduction of high-frequency PVC welding, inflatables became a realistic proposition. The 'Blow Chair' designed in 1967 in Milan by Scolari, Lomazzi, D'Urbino and de Pas, was the first commercially produced inflatable armchair since the 'Elliot' of 1942, and the first totally inflatable chair. Whereas the 'Numax' had a frame for use inside the house, the 'Blow Chair' totally abandoned legs. In addition, it was completely transparent, making visible every structural detail in its design.

A great advantage of air-supported structures is their instant but impermanent availability; but their transience has not been a hindrance to grandeur, as can be seen

above
The *RFD Inflatable Chair, c.* 1947. (RFD Inflatables Ltd, Godalming, Surrey)

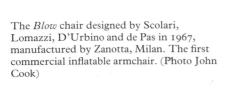

The *Blow* chair designed by Scolari, Lomazzi, D'Urbino and de Pas in 1967, manufactured by Zanotta, Milan. The first commercial inflatable armchair. (Photo John Cook)

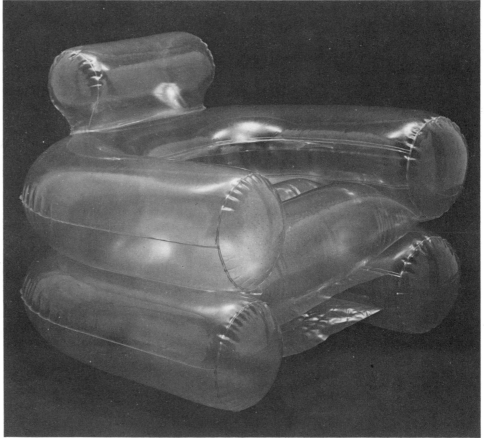

by two pneumatic landmarks of great size: the BUC Cupola and the Fuji Pavilion. The former is the more recent, and when built in 1972 for the Compagnie Française des Pétroles at Yvelines, France, it was the world's largest inflatable structure. Based on a simple plan, one hundred metres square and fifteen metres high at the centre, it looks like a domed skylight over some subterranean world. It is constructed from a PVC-coated 'Trevira' outer skin held down by hooks to a lattice of double steel cables anchored to concrete foundations. The air inside which supports the structure is renewed every one and a half hours.

But perhaps the most spectacular inflatable in terms of form and colour was the Fuji Group Pavilion at the Osaka Expo in 1970, which was also the world's largest pneumatic structure when it was first built. It has a circular plan fifty metres in diameter. Sixteen airbeams are anchored around the circumference, each beam four metres in diameter, and seventy-two metres long, consequently forcing the beams to bulge higher at each end, where they are pressed closer together, whereas the centre beam forms a semi-circular arch. The ribs are held together by horizontal straps. It took only thirty minutes to inflate, and the same time to collapse.

Inflatable chair designed by William H. Miller, *c.* 1944, manufactured by the Gallowhur Chemical Corporation, New York. A doughnut-shaped PVC tyre is held by netting onto a timber frame. (Museum of Modern Art, New York; gift of Gallowhur Chemical Corporation)

79

Fuji Group Pavilion, Osaka Expo 1970, designed by Yutaka Murata and Mamoru Kawaguchi. One of the most colourful inflatables ever, recalling the 'Michelin Man and other anatomical wonders' (Charles Jencks). Constructed with air beams made of two layers of PVA-impregnated canvas, the inner surface insulated with PVC, the outer skin coated with Du Pont's 'Hypalon' synthetic rubber which is easily applied for repairs, and can be pigmented with bright colours. (Photo Mary Allwood. *Architectural Design*)

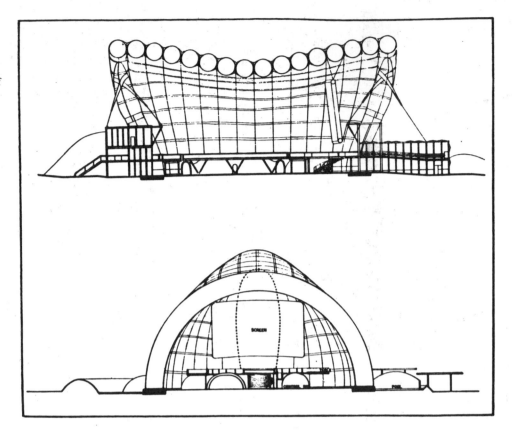

PVC in Art

Just as PVC introduced new possibilities into furniture design and architecture, its potential was exploited on a different level by various artists. Claes Oldenburg was the first artist to upset the convention that artistic materials should be solid and permanent. Around 1960 he started to make bulging shapes of shiny PVC stuffed with kapok, later called 'Soft Art'. His sculptures took the form of gargantuan soft food, such as his *Bacon, Lettuce and Tomato* sandwich of 1963, or domestic hardware, such as limp light-switches or the famous *Soft Typewriter*, also made in 1963, which became the symbol of the Soft Art Movement.

At the present time Soft Art has reached a level of 'hyper-realism' with the plastic polyurethane 'Nature Carpets' of Piero Gilardi.

Expanded PVC

Like many other plastics PVC can be produced as a cellular foam both rigid and flexible in varying densities. It is not used as much as polyurethane foam or rubber latex foam, as it is more expensive to manufacture.

Plastic foam is manufactured with two types of structure: open-cell, with its cells interconnecting like a sponge, similar to flexible polyether foam or foam rubber, and closed-cell, a rigid honeycomb of non-communicating bubbles. PVC foam is moulded in sheets as it cannot be foamed *in situ*. Rigid closed-cell PVC is obtained when a blowing agent in the form of a powder is blended with the vinyl resin compound. It produces its own heat in the mould, or else heat is applied, and releases nitrogen which forms bubbles, causing the molten compound to expand. Rigid sheet foam of this type is used as a strong insulating material, similar to urea and polyurethane foams. Some grades are as hard as steel and are used to insulate submarines. Not counting structural foams, PVC foam is the strongest of all the rigid foams.

Cellular rigid PVC is enjoying a boom at the present time. In America the market for moulded architraves, skirtings, cornices and even coffins began to increase in 1972, and between 1971 and 1973 the sales of PVC foam profiles in America leapt from nil to 20,000 tonnes per annum. In Britain before the oil crisis of 1973 BIP reported that 'Beetle' cellular rigid PVC extrusions were poised to take over the market for timber mouldings. Although the price of oil quadrupled putting up the

Bedroom by Claes Oldenburg, constructed for the exhibition 'Four Environments' by the New Realists, 1964, Sidney Janis Gallery, New York. Apparently based on his memory of a West Coast motel, Oldenburg has made everything disconcertingly synthetic and vulgar. It is almost a parody of the well-intentioned Pierce Foundation room (see page 77). (Photo Geoffrey Clements)

BUC Cupola, designed and built in 1972 by Air Structure G.A. Sceaux, France. An air-supported structure with a skin of PVC-coated 'Trevira'. (*Building Design*)

price of PVC, by 1976 sales had returned to the 1972-3 level. PVC foam can be easily sawn, nailed, stapled and glued by any home decorator like traditional wood, and is ready-finished. Only a top-coat is necessary if it is to be painted. It is also a versatile material for building prototypes. In 1974 'Plasticell' protected by a layer of GRP was used for constructing a nose section for the Formula 1 John Player Special racing car, and it has been used frequently to build yacht hulls with a double skin of GRP over a foam core.

Although a ton of PVC costs more than a ton of wood, the wastage of wood during processing to an end product is reckoned to be as high as 50 per cent. With an expansion in world trade, which could overload timber supplies, PVC 'wood' will probably become an economical proposition.

The main application of flexible open-cell vinyl is upholstery, where the material needs to breathe. The PVC forms a very soft and resilient skin over a reinforcing backing of stretch nylon jersey.

70-foot long boathull made of expanded closed-cell rigid 'Plasticell' PVC, skinned with GRP, over a timber frame. The sandwich construction provides thermal insulation, lightness, strength and moisture-resistance. (BTR Industries Ltd)

Extruded PVC Monofilaments

PVC monofilaments can be found in the construction of artificial golf courses and ski slopes, where, unlike real situations which deteriorate, all competitors are given the same, if unnatural, conditions. The resilient toughness of PVC monofilaments is due to the orientation process (see section on processes, p. 170) which produces fibres which will never deform but which spring back with what is called 'shape memory', always returning to their originally moulded form.

Polyvinyl Acetate (PVAc) *thermoplastic*

PVAc was known in the thirties but it was not much used until the forties. Although never equalling the use of urea-formaldhyde and phenolic, its greatest application was as a cement and wood adhesive. It is still used as a milky-white glue for wood and paper and as a contact adhesive, although its main application is in the field of water-based emulsion paints. It is a rubbery clinging glue. In the forties it was also used as a coating to waterproof fabrics for raincoats and sheeting. With an added filler, PVAc can be moulded into tiles, and has been used to replace the chicle in chewing gum.

By making slight changes to the side groups on the polymer chain, PVAc can be made into polyvinyl alcohol, a vinyl with a higher softening point which is used for gaskets and chemical-resistant containers. Further modification of PVAc produces polyvinyl formal (a wood adhesive), polyvinyl butyral (a waterproof coating and safety glass laminate) and polyvinyl acetal.

Polyvinylidene Chloride Polymers (PVDC) *thermoplastic*

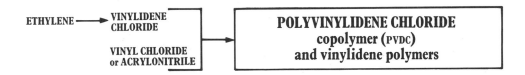

The first vinyl chloride fibre was made by the Germans into tough air force clothing and called 'Pe Ce'. The Dow Chemical Company in America in 1940 was the first to produce polyvinylidene chloride (PVDC) commercially (from the monomer vinylidene chloride and cyanide) under the trade-name 'Saran', a name later used generically to describe any polyvinylidene chloride plastics.

PVDC is a chemically resistant, non-flammable and non-toxic material which can be injection-moulded, compression-moulded and extruded. Its chief use is as extruded yarns and filaments for weaving upholstery on public transport, deck chairs and protective clothing, and it is also used in dental floss, fencing, fishing nets and industrial filters. In the forties it was widely used industrially: extruded in tube form it was common in laboratories, and as an early replacement for cast-iron plumbing.

Vinyl trade-names include:
UK: Breon (B.P. Chemicals), Corvic, Darvic, Flovic and Welvic (ICI), Beetle, Cobex, Extrudex, Velbex and Vybak (BIP), Fablon (Commercial Plastics), Plasticell (BTR Industries), Portex (Portland Plastics), Numax (RFD-GQ Ltd), Velbex (BX Plastics). Floor tiles: Amtico, Marleyflor and Polyflor. UPVC underground drainage systems: Hunter, Marley, Osma and Plastidrain. Leathercloth: Ambla, Cirrus, Duflex, Everflex and Vynide
USA: Elvax (Du Pont), Geon and Koroseal (B. F. Goodrich Chemical Co.), Naugahyde (Uniroyal), Opalon and Ultron (Monsanto), Vinyon (Bakelite/American Viscose), Vitrolac (RCA records), Vinylite (Union Carbon and Carbide Corporation)
GERMANY: Vinoflex (BASF)

Saran trade-names include:
UK: BX Saran (Bakelite Xylonite), Viclan (ICI)
USA: Saran (Dow Chemical Co.), Velon (Firestone Plastics Co.)
ITALY: Ravinil (ANIC)

POLYSTYRENE *thermoplastic*

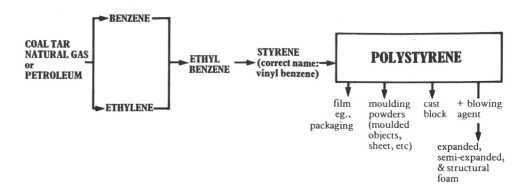

Polystyrene is the fourth ranked polymer, after polypropylene, polythene and PVC, in volume production. The styrene monomer for making polystyrene is obtained from benzene and ethylene, both petrochemical feedstocks. At present polystyrene is entirely dependent on this source, and consequently affected by oil prices.

The styrene molecule is basically an ethylene molecule with one of its four hydrogen atoms replaced with a benzene ring:

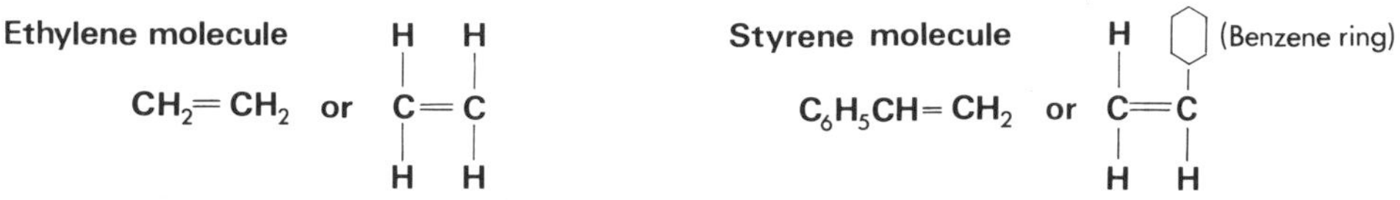

Ethylene molecule

$$CH_2 = CH_2 \quad \text{or}$$

Styrene molecule

$$C_6H_5CH = CH_2 \quad \text{or}$$

(Benzene ring)

Vinyl benzene had been successfully polymerized in England in 1911 by Mathews, who recommended it as a substitute for celluloid, but it was not until Hermann Staudinger's work at Freiburg from 1926 that serious interest was aroused. The German company BASF was the first to pioneer the commercial production of polystyrene in 1929 using secret processes. Not until around 1935 did reports and pictures reach America and Britain of a new German 'poly-styrol' material called 'Trolitul', manufactured in clear and patterned sheet by Eckert and Ziegler, part of I. G. Farbendindustrie. Polystyrene was first made in America in 1937, when 'Styron' by Dow, and Bakelite polystyrene by Bakelite, were commissioned by the government to make electrical insulations.

In post-war Britain and America 'Styron' houseware became fashionable. 'Beautiful and practical', it quickly appeared as picnic sets, dinnerware, disposable cutlery, jugs and serving dishes. The advertisements proclaimed: 'Just like magic! . . . Styron is the right plastic for a host of functional duties in today's active living.' Its attractive colours and crystal clarity gave a jewel-like and luxurious impression at cheap prices, but it was soon discovered that polystyrene cracked and crazed.

Its high-frequency electrical insulating property made polystyrene the best material available in the forties for cabinets, for radios and televisions (formerly a thermoset stronghold), storage battery containers and insulation for wiring in aircraft. It was tough enough to protect cables that had to be pulled through deserts or ice.

Another valuable property is optical clarity and, like polymethyl methacrylate (acrylic), it can carry light around corners. This, and its low moisture absorption, have made it useful to doctors and dentists for visual probes.

When heated, polystyrene flows exceedingly well and is therefore easy to mould into intricate shapes with accurate dimensions for such objects as maps for the blind. Thousands of decorative novelties are made in transparent polystyrene, sometimes

84

containing metal flakes in the resin, such as boxes, toys, costume jewellery, spectacle frames and ballpoint pens. However, polystyrene is one of the more brittle plastics, although this brittleness can be overcome by orientation (in which the molecules in the polymer are stretched while hot), as is done in making film for blister packaging. To counteract crazing a toughened form of polystyrene, HIPS (high-impact polystyrene), has been developed. Other styrene-based polymers include SAN and ABS. Polystyrene is converted by the standard thermoplastic processes: injection moulding, extrusion, thermoforming, rotational and blow moulding.

High-Impact Polystyrene (HIPS) (also called 'toughened' or 'impact' poly-styrene)

To improve impact and heat resistance, polystyrene is compounded with synthetic rubber (butadiene) in the manufacture of toys, furniture, vacuum cleaners and cameras, all of which are likely to suffer damage in use. Food containers for dairy produce are injection-moulded with very thin walls to take advantage of the increased impact resistance and refrigerator liners are made that do not crack. HIPS sheet is used in vacuum forming.

HIPS has replaced timber and metal in reproduction antique furniture, where such a degree of sophistication has been reached that close inspection reveals the sanding and chisel marks of the original wooden pattern. This type of furniture is immensely practical: minimum maintenance is required, it is light and tough and, unlike timber,

Olympic All Polymer Chair, 1968, moulded in Lustrex by Plastic Industries Inc., Tennessee, pioneers of the first solid high-impact polystyrene components with which to assemble a range of furniture. There is no limit to the styles that can be reproduced.

opposite
Mushroom Range designed by Maurice Burke in 1965, for Arkana Ltd, and produced in 1967 as the first rotational-moulded furniture in Britain. Moulded in B.P. high-impact polystyrene with a tough butyrate finish, the double-shell seat and the hollow bases of tables and chairs are filled with polyurethane foam while the table has additional ballast. The rounded forms are typical of rotational moulding. Though comfortable, there is something inelegant about the design which removes it from the category of formal dining furniture to fun furniture. (Hollen Inc.)

Folding armchair designed by Fred Scott in 1974, manufactured by Hille International Ltd. Moulded by S. A. Synfina, Belgium, on a two-minute cycle. High-impact polystyrene seat and back on tubular metal frame. The arm/back support is reinforced with a metal plate achieving an impressive 22.8cm (9in.) cantilever, and can support the weight of a 16 stone man. The seat is deeply dished and only 0.48cm ($\frac{3}{16}$in.) at the centre. The first use of this material in modern-style furniture. A pad can be attached to the back rest to counteract comfort problems due to production changes.

not affected by damp. It is highly popular in the USA: in 1964 there were twelve presses supplying approximately 1,300 tons of furniture parts: in 1969, 350 presses supplied 150,000 tons. In Britain reproduction furniture of this type is rarely seen.

Styrene-Acrylonitrile Copolymer (SAN) *thermoplastic*

The low chemical resistance of polystyrene can be improved by copolymerizing it with acrylonitrile, producing styrene-acrylonitrile copolymer, commonly known as SAN. It still looks and feels very much like polystyrene but is much stronger and has a high-gloss finish; in its transparent form it is often used for liquidizer bowls, tableware, and record player covers. It is also used for many types of prepackaged food containers, and a large variety of handles and knobs. The keys of the Hammond Electrical Organ, originally ivory and ebony, and then urea and phenolic in the thirties, are now injection-moulded styrene-acrylonitrile.

A more recent and tougher version of SAN has been developed, known as ASA terpolymer, which is styrene-acrylonitrile copolymer modified with acrylic elastomer. It is suitable for outdoor use such as guttering and stadium seating, and was the material used for Verner Panton's stacking chair and for the 'Mono Chair' designed and developed by Dominic Michaelis and Simon Fussell.

Semi-Expanded and Expanded Polystyrene (XPS)

Semi-expanded polystyrene is loose and granular, familiar from the white pearl-like balls used as stuffing inside 'sack' chairs. Expanded polystyrene is used for white insulation tiles, mouldings for the do-it-yourself market, shaped boxes for packaging electrical goods, foam chips used in insulating and packaging, and squeaky pearlized egg cartons. Polystyrene is excellent for insulating and packaging; it absorbs no moisture and therefore does not rot; it is firm yet flexible.

Semi-expanded polystyrene is produced by impregnating polystyrene granules with a blowing agent (pentane) and heating with steam to blow them up to several

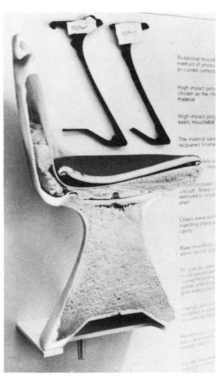

Cross-section of the *Mushroom* chair.
(Council of Industrial Design, London)

87

The *Mono Chair*, part of a range of chairs with or without arms and upholstery. Designed and developed for Interstore by Dominic Michaelis and Simon Fussell in 'Luran s' (ASA terpolymer by BASF, Germany). (Interstore)

times their original size, producing a 'pop-corn' effect.

In making polystyrene chair and settee shells, matured semi-expanded beads are placed inside matched male and female moulds which are then steam-heated to expand and fuse the beads into one fully expanded foamed mass. The shells are always upholstered, since the surface of the foam has no strength and can easily be dented. The strength of the shells as furniture carcassing lies in their compound curves, like eggshells, but since 1974 there have been reports of failure, typically of sections breaking off.

Since expanded polystyrene can be machined and cut with hot wires or knives, it has often been used by sculptors. Jean Dubuffet shapes and then paints his reliefs; Gino Marotta in Italy carved an 'Artificial Cave' like a giant insulating womb. The surface of polystyrene can be filled and sprayed with paint, provided it is not an oil-based or cellulose-based paint, which will dissolve the polystyrene.

Polystyrene Structural Foam (PS SF)

The recent development of structural foams has brought a new image to plastics. Structural foam plastics are an important step in the gradual replacement of traditional materials like wood and metal with man-made chemical substances. Polystyrene, like other plastics such as polythene, polyurethane, polypropylene and polycarbonate, can be manufactured in a structural foam grade, that is, a rigid plastic with a strong skin over a foamed core. These foams are thin in section, but are very tough and shock-resistant and look like ordinary unfoamed plastics. The advantages of these materials are their low cost, light weight, impact and vibration resistance, mould detailing, integrally moulded inserts, and small number of parts. Structural foam plastics moulded at low pressures have a characteristic swirl pattern on the

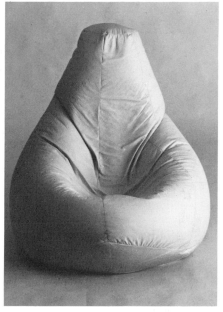

Sacco Chair designed by Gatti, Paolini and Teodoro in 1968, manufactured by Zanotta, Milan. 90 × 128cm (35 × 50in.) high. It is a bag filled with millions of semi-expanded polystyrene beads 3mm diameter. The first sack of its kind, it disproves the theory that one chair cannot suit both men and women, since being only half-filled it adapts itself to any size or shape of body.

The mould for the *Mono Chair* being cut from a steel block by a cutter following dimensions recorded from the model. This mould cost £50,000 in 1973, 'a very expensive piece of abstract sculpture indeed' (Michaelis), which precludes further modifications for other markets. (Photo by courtesy of Dominic Michaelis)

surface, and are used either as reproductions of timber, or beneath a layer of paint as audio-visual and machine housings, or where visual appeal is not a priority, as in beer crates and industrial containers.

In 1973 the first plastics piano was designed by Deschamps-Mills Associates. Their brief was to design a reasonably priced beginner's piano. Polystyrene structural foam proved a suitable substitute for the expensive traditional hardwoods such as mahogany and pecan. Not only is structural foam cheaper but it feels and

The first plastics piano designed by
Deschamps-Mills Associates in 1973 for the
Estey Piano Corporation. Front and back
view of side panel moulded in Union
Carbide's high impact polystyrene structural
foam by the Cashiers Plastics Corporation,
North Carolina. (Cashiers Plastics
Corporation)

sounds like wood, and in addition vibrates less. The entire piano can be easily dismantled with a screwdriver.

The long, flat surfaces of the piano are made of hardboard and particle board veneered with timber-grained vinyl foil. The sounding boards are solid spruce and the bridges are maple, but the sculptured end panels, legs, music rack, key blocks and filler plates are moulded in Union Carbide's high-impact polystyrene structural foam which was first introduced in 1966. Structural foam mouldings can be made under low pressure, and consequently moulds can be made of relatively cheap materials such as aluminium, or, as in this case, beryllium-copper.

Polystyrene structural foam is also used for keyboard instruments by Wurlitzer and Hammond Organ for their latest electronic organs. The new Wurlitzer 'Sprite' models are constructed almost entirely of injection-moulded polystyrene structural foam components, and include other plastics, such as polypropylene, in the form of integral hinges.

UK: Bextrene (BXL), BP Polystyrene (BP Chemicals), Carinex (Shell Chemicals), Distrene and Bexfoam (BX Plastics), Transpex 2 (ICI)
USA: Lustrex (Monsanto), Styrofoam, Styron, Styrex and Styrite (Dow Chemical Co.), Union Carbide Structural Foam

GERMANY: Hostyren (HIPS, Hoechst), Luran (BASF), Polystyrol (HIPS, and HIPS foam, BASF), Styropor and Styrofil (BASF)
ITALY: Edistir (HIPS) and Edistir RV (HIPS foam, Montedison)
SAN trade-names include:
USA: Lustran (Monsanto), Tyril (Dow/Distrene)
GERMANY: Luran S (ASA by BASF)
ITALY: Kostil (Montedison)

POLYMETHYL METHACRYLATE (PMMA; Acrylic) *thermoplastic*

Polymethyl methacrylate is commonly known as 'acrylic' or by trade-names such as 'Perspex', 'Plexiglas' and 'Lucite'. Acrylic resins come in two varieties, polymethyl methacrylate and polymethyl acrylate.

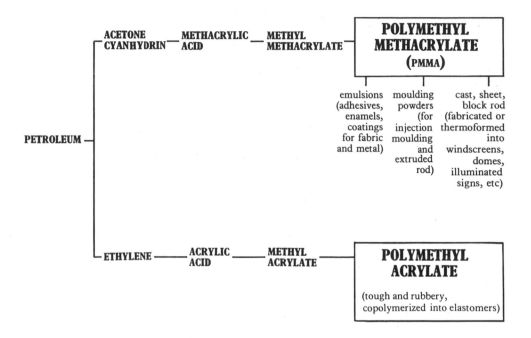

A glass-like acrylic resin was first formed by Rudolph Fittig in 1877 and developed in Darmstadt by the German chemist Dr Otto Röhm who polymerized methyl methacrylate in 1901-2. However, economic commercial production of the acrylic monomer was fraught with difficulties until the firm of Röhm and Haas in 1928 succeeded in producing polymethyl methacrylate (PMMA) commercially in a range of clear soft resins, which were initially applied as coatings. Like many chemists before him, Röhm had created a sticky plastic resin and looked for ways of making it into a useful adhesive. As with other plastics, further chemical reactions later turned the soft polymer into a mouldable rigid plastics material.

The production of the harder form of acrylic is considered a British success, achieved in 1934 by two chemists at ICI, Rowland Hill and John Crawford. By 1936 commercial acrylic cast sheet was on the market under the trade-name 'Perspex', and around the same date the Germans brought out 'Plexiglas' acrylic sheet, in time to supply the Luftwaffe aircraft.

Acrylic sheet is cast between plates of glass which give it its smooth, high polish. Blocks and rods can be cast and extruded. As a replacement for glass, it is shatter-proof and possesses excellent clarity, which made it ideal during the war for use in aircraft: transparent nose and tail sections, cockpit canopies, gun turrets, observation blisters, windshields, windows, navigation light farings and bomb sight hatches. Many ex-RAF men still possess model fighter planes and novelties they made from broken acrylic aircraft parts whilst grounded.

In 1937 ICI brought out powder and granule forms of acrylic suitable for making

Cockpit section of an Avro Lancaster BI, with nose, cockpit hood, blisters and front gun turret blow-moulded in 'Perspex'. (Imperial War Museum, London)

The streamlined observation coach at the rear of the 'Coronation Scot' LNER express built in 1937 to celebrate the coronation of King George VI. The 'beavertail' windows were thermoformed from 'Perspex' by the Triplex Safety Glass Co. (Crown Copyright. National Railway Museum, York)

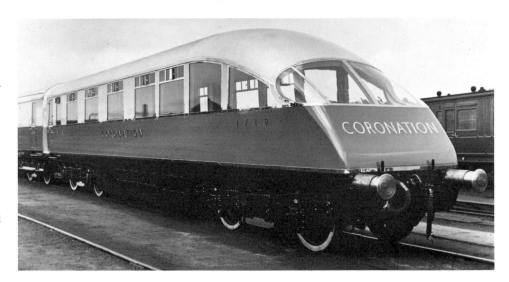

below
Pontiac with its body, wings and bonnet shaped in 'Plexiglas' at New York's World Fair, 1939. Acrylic has always been used by designers working out designs for machine parts. The Motor Show in London 1938 had revealed a Hillman Minx behind 'Perspex' panels attracting the crowds with the slogan 'Watch your friend double-declutch'. (Pontiac Motor Co.)

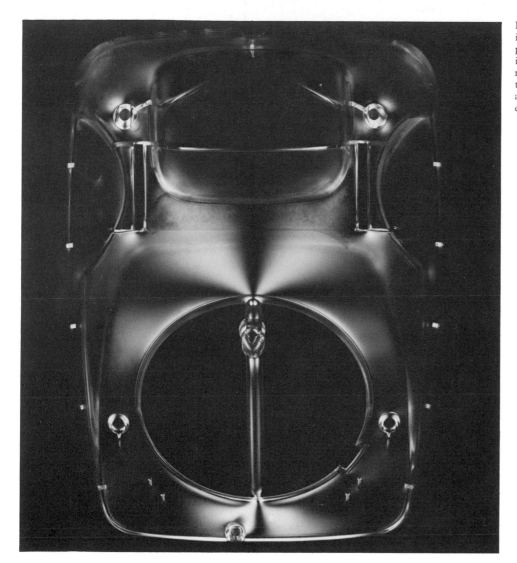

Experimental moulding of a telephone cover in 'Diakon' (ICI acrylic polymer), photographed under polarized light which indicates the distribution of stress in the moulding through strain colours (colours of the spectrum). Cast phenolic and cellulose acetate have also been used for photo-electric analysis. (ICI)

transparent or translucent mouldings under the trade-name 'Diakon'. It could be compression-moulded, injection-moulded and cast. The year 1937 was also important for dentists: ICI introduced an acrylic for dentures—at last a new colourless, easily workable substance, and still the main material for that purpose. Also in 1937 Du Pont produced acrylic sheet and acrylic in various forms suitable for moulding under the trade-name 'Lucite'.

As 'Perspex' decomposes at high temperatures, the development of 'Diakon' made possible the moulding of all kinds of solid objects: cutlery handles, door knobs, salad servers and bathroom fittings, jewelry and decorative articles. Today's coloured 'Perspex' was not developed until after the war.

The Optical Properties of Acrylic

One of the most attractive properties of acrylic is its optical clarity, making it a popular material when display is required. 'Perspex' boxes are now often used to display objects in museums without distortion, and the J. Paul Getty Museum in Malibu, California, protects its art works from the dangers of ultra-violet light with special lenses moulded in 'Plexiglas UVA-7' by Röhm and Haas. In the forties lenses were made which took advantage of the refractive property of acrylic for enlarging; cast acrylic blocks fixed to dials and gauges made for easy reading. Today many spectacle lenses are acrylic. Contact lenses have been greatly improved by acrylic; they were originally made of glass and had to be either blown or very

Ceiling composed of thermoformed 'Lucite' in the Patent Leather Lounge, St Francis Hotel, San Francisco. Designed by Timothy Pfleuger and completed in 1939, it is one of the most impressive early applications of acrylic as light diffuser. (Pfleuger Architects)

carefully hand-ground. Soft contact lenses appeared around 1972 and were made of hydrophilic acrylic (chemical name: hydroxy ethylmethyl methacrylate), a resin developed by the Czechoslovak Academy of Science. Marketed in the USA as 'Hydron', this material is moulded into valves for artificial hearts and can also be found as a surface coating on industrial masks and ski goggles where it forms a non-fogging finish, absorbing moisture. This property is also put to use in its application as a coating on concrete, brick walls and boat hulls, where like silicone it lets moisture out but not in.

Like polystyrene, acrylic carries light, even round bends, and is put to great use in medicine and dentistry to 'pipe' light along probes. In an early application of fibre optics, images were scrambled by bundles of 'Perspex' rods before transmission in wartime operations. Nowadays lamps can be bought looking like sea-anemones, waving multi-fibred tentacles, the end of each one a pin-prick of 'piped' light. They are made of bundles of filaments connected to a hidden light source.

Hundreds of different lampshades and diffusers have been moulded in acrylic by Italian firms such as Artemide, Kartell and Guzzini, varying from ordinary spheres and cubes to the more imaginative and improbable. Marcello Siard's floor lamp for Kartell was designed to be pushed around on wheels like a pram. Vico Magistretti's floor lamp for Artemide resembles a wavy screen of luminous satin, and is similar to a design by the Italian sculptor Sante Monachesi—an acrylic sheet draped like fabric, standing like a petrified curtain without a curtain rail.

Acrylic Furniture and Jewelry

Users of acrylic sheet and rod for furniture fall into two camps: either they imitate the forms created by other materials, or else they exploit acrylic's properties in an imaginative and inspired way.

As an example of the former, reproduction acrylic furniture in the USA reached a remarkable level of verisimilitude in the forties, only surpassed by recent work in high-impact polystyrene and polyurethane structural foam. It was usually assembled in the same way as the original, with pieces hand-shaped and screwed together. This quite ignores the basic properties of the material, which is best used in the form of curved shapes that exploit its tensile strength. Designers since then have likewise often neglected the specific plastic properties of acrylic which have to be respected if comfort and originality are to be achieved. Acrylic seats can be made by thermoforming sheet into a curved shape and clipping this to a frame, as in David Colwell's 'Contour' chair of 1968. A cheaper way in which the sheet can be curved to form furniture, which does not involve any heat or moulding equipment, is the simple curving and fixing of cold flat sheet. Anthony Hooper's 'Bendit 2001' chair designed in 1970 reaches the customer as a flat sheet which is then neatly bolted on each side into a curved steel frame. The design is basic enough to suit other flexible sheet materials such as polythene and polypropylene.

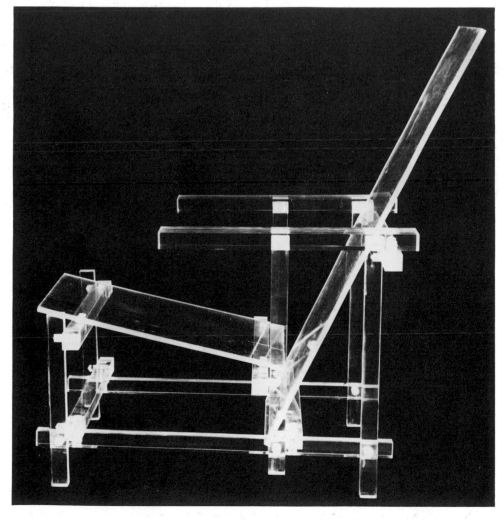

Acrylic facsimile of Rietveld's 'Red-Blue Chair' of 1917-18 designed *c.* 1968 by Kho Liang Le. Acrylic negates associations of material, colour and texture, in keeping with Rietveld's De Stijl philosophy that 'No part dominates or is subordinate to the others. In this way, the whole stands freely and clearly in space, and the form stands out from the material.' In making the skeleton transparent, the engineering forces in the chair are made visible. Property of the Bruynzeel Factories at Krommenie. (*Design*)

opposite
First acrylic heart, and Italy's first man-made heart. Designed by the bio-engineer Robert Bosio in 1965; acrylic with silicone membrane. (*Materie Plastiche ed Elastomeri*)

The early seventies saw a plague of acrylic nesting tables, occasional tables, magazine racks, lamps and furniture in all shapes and sizes which unfortunately often ended up scratched and dusty, and television sets which displayed their internal wiring systems inside transparent globes.

In 1967 ICI held a promotion exhibition of student work at the Royal College of Art in London entitled 'Prospex 67' to illustrate the multifarious applications of Perspex. The exhibition had a strong impact on British designers. Particularly influential was Suzanne Fry's colourful Perspex jewelry, and before long her large laminated rings were to be seen in any high street. Nowadays acrylic is completely accepted as a jewelry material. It possesses the beauty of glass with the advantages of the freedom offered by a thermoplastic material—ease of shaping and texturing, a wide range of colours, as well as transparent and translucent, and the ability to laminate different colours together producing an agate-like effect. This material is now paired quite readily with gold, silver and precious stones. David Watkins in particular has evolved an individualistic use of the material. His pieces are made of curved rods, sand-blasted and held together with silver pins and bands.

The Versatility of Acrylic

Acrylic can be found in innumerable applications, but especially where it comes into contact with water, chemicals or the human body. Since it is warm and smooth to the touch, it is used for hand-made door handles, finger plates, escutcheons and cupboard knobs, as a poor man's version of alabaster, jade, crystal and marbled onyx, in the same way that cast phenolic was once used; it is also favoured for solid 'marble' handbasins, transparent tinted bathroom accessories, children's thermoformed cutlery, as well as for jewelry, and hairbrush and bag handles.

La Vasca Tonda (The Round Tub), 17 × 17 × 21 metres (56 × 56 × 70ft), part of the Aquarius range of bathroom units, designed by Fabio Lenci, manufactured by Teuco, Italy. Injection-moulded in glass-reinforced 'Vendril' and 'Vitredil' acrylic supplied by Montedison. More a bathroom than a bath, it is very streamlined, very compact, and very expensive. (Victor Mann and Co. Ltd)

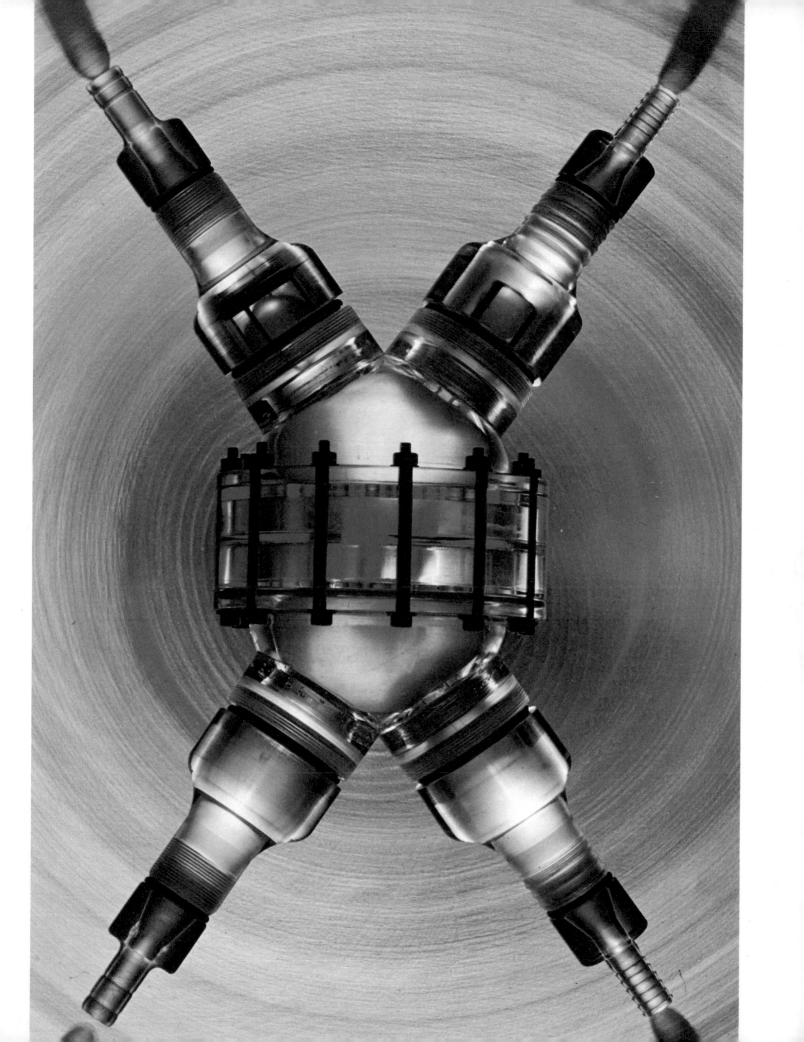

In maternity hospitals most modern wards now cradle their newborn babies in transparent acrylic cribs, an improvement on the old steel models. In surgery certain acrylic prosthetic parts are acceptable to the human body: tubing, heart valves, esophagi, tracheae and even eyeballs, and foamed artificial breasts.

The most extravagant novelty use of acrylic must surely be the 200-dollar toilet seat and lid manufactured by the 'most reputable leather and novelty business' in Salt Lake City. It is made of cast transparent acrylic with coins embedded into it, which will, of course, increase in value. Apparently for its next design the company planned a toilet seat with barbed wire embedded in it.

Acrylic has been used as an artistic medium, notably in Kinetic Art, which has exploited its optical qualities. Since Gabo and Pevsner first used celluloid sheet in their constructivist sculptures in the twenties, artists have experimented with plastic materials. Acrylic is particularly suited to the optical properties of Kinetic Art as it can create moiré patterns on printed or engraved sheets, whether as an optical illusion, an actual moving object, or an effect produced by the spectator moving in front of the work. It is also exploited for its ability to transmit light, for the effects achieved by light, water and colour refracted or laminated into the material, and for its flexibility in moulding and shaping by hand.

In the early sixties acrylic paints played an integral part in the development of the Hard Edge and Colour-Field schools of painting. Artists such as Kenneth Noland and Morris Louis stain unprimed canvases with washes of acrylic diluted in oil or water. Acrylic paints are not as easy to manipulate as oil paints. They dry very quickly, but it is possible to achieve very intense colours.

As we have seen, acrylic has many advantages over glass. It is approximately half the weight; it is weather-resistant; it is more difficult to break or vandalize; it has flexibility under stresses that would shatter glass; it is easily portable and demountable; and it can be hand-shaped with heat. However, the disaster at the Summerland Solarium on the Isle of Man in 1973 has shown that great care must be taken in applying plastics materials. For despite the advantages listed above, acrylic is a combustible material—it burns within minutes, and is not self-extinguishing like many plastics. As the official report states (*The Architect's Journal*, 29 May 1974), fifty people died 'needlessly' when £63,000 of acrylic melted and dripped in the inferno. Perhaps the design had been inspired by Buckminster Fuller's geodesic US Pavilion at the Montreal Expo in 1967; Fuller had also used domes of acrylic. But despite its strict adherence to fire codes, that too was to be destroyed by fire in May 1976.

Foamed Polymethyl Methacrylate

Recently ICI has developed a rigid foam with an integral skin. The process involves compressing a slab of expanded acrylic foam between heated platens which shape it and give it its hard skin. Where it is almost totally compressed the material goes translucent. So far, it has been marketed in the pattern of standard timber external doors. ICI has produced doors with regular depressed panels simulating inset glass panes. It is very tough with a superb glossy finish, the sort of finish not possible with painted wood, except with a disproportionate amount of labour and the right conditions. But being self-coloured it does not need painting and it is weather-resistant and rot-proof. The foam can easily be worked with standard joinery tools and takes screws well.

Polymethyl Methacrylate Elastomers

Elastomers are rubber-like materials made from synthetic plastics, produced in a variety of forms (see section on Synthetic Rubber, p. 27). Acrylic elastomers are not all that new, having been patented first in Germany in 1912 by Dr Otto Röhm. Later the American Cyanamid Co., now Cyanamid International, developed 'Cyanacryl', and the Thiokol Chemical Co. introduced 'Thiacryl'.

'Cyanacryl' acrylic elastomers are used to replace the more expensive silicone rubbers and the nitrile rubbers in the manufacture of car parts. 'Cyanacryl' remains stable in high and low temperatures and resists hot oils and the effects of weather.

Polyacrylates have been used as adhesives for a long time. One developed for the Apollo Moon Missions by NASA is so powerful that only one drop is needed to make a repair and it sets in less than a minute. It is used in dentistry to repair acrylic dental plates and teeth, although it is not used directly in the mouth. In Vietnam, temporary repairs to vascular wounds were carried out with cyanoacrylate spray in emergency surgery. If a patient was likely to die from a haemorrhage, damaged veins and arteries could be quickly stuck together to stop the bleeding.

Acrylic (Acrylonitrile) Fibres

Acrylic fibres are an important group of plastic fibres, and sales have boomed. They are high performance fibres which can be dyed with intense colours, and are highly resistant to chemicals. Their use, along with that of other synthetic fibres, has contributed to the scarcity of moths in wardrobes today.

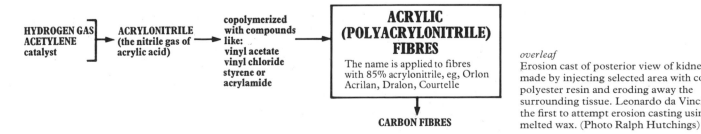

HYDROGEN GAS / ACETYLENE catalyst → ACRYLONITRILE (the nitrile gas of acrylic acid) → copolymerized with compounds like: vinyl acetate vinyl chloride styrene or acrylamide → **ACRYLIC (POLYACRYLONITRILE) FIBRES** The name is applied to fibres with 85% acrylonitrile, eg, Orlon Acrilan, Dralon, Courtelle → **CARBON FIBRES**

overleaf
Erosion cast of posterior view of kidneys made by injecting selected area with coloured polyester resin and eroding away the surrounding tissue. Leonardo da Vinci was the first to attempt erosion casting using melted wax. (Photo Ralph Hutchings)

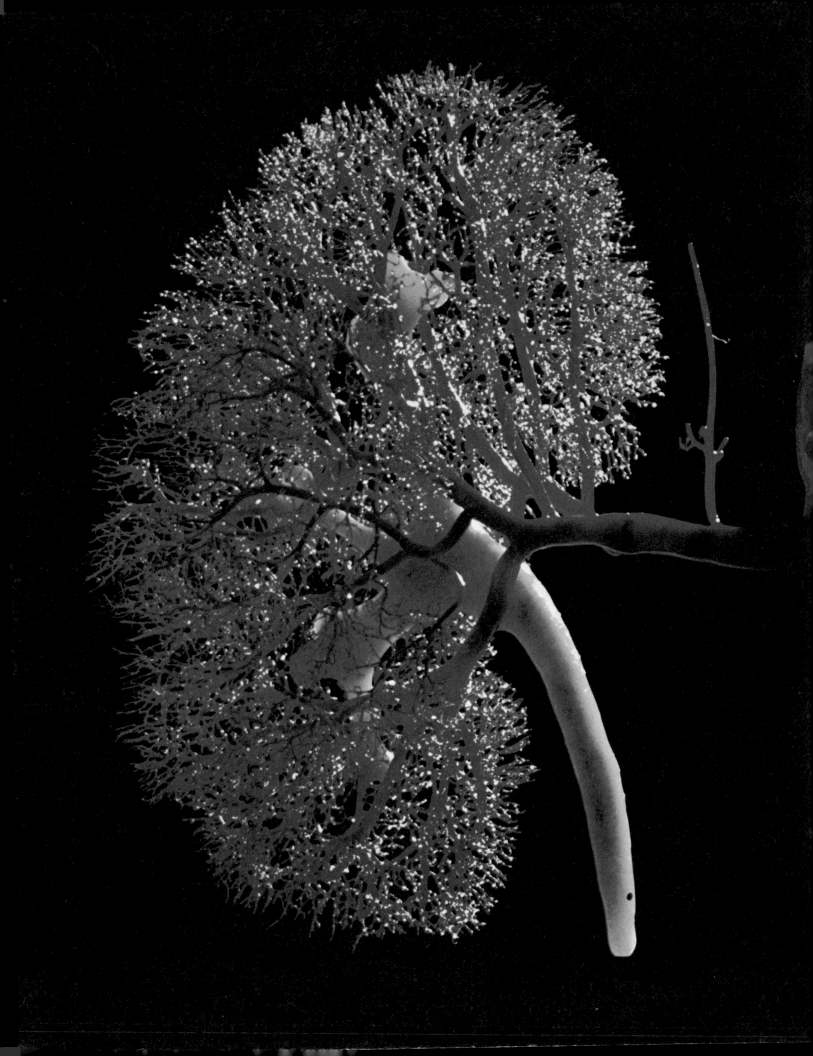

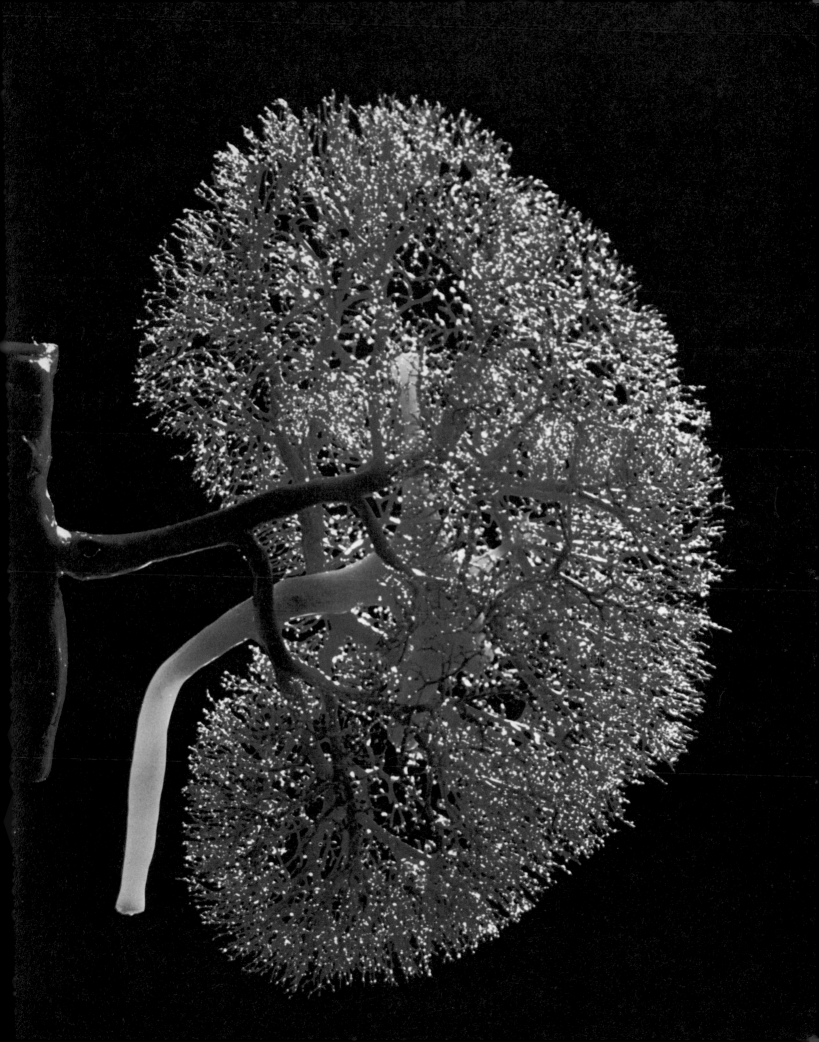

Copolymerization with acrylonitrile increases the chemical resistance of a polymer, whether acrylic fibre, styrene acrylonitrile, or a synthetic rubber like polybutadiene acrylonitrile. The first acrylic fibre was made by Du Pont in South Carolina in 1950 under the trade-name 'Orlon', but others soon followed: 'Acrilan' (Chemstrand Corporation), 'Dralon' (Bayer), and 'Courtelle', the first British version, produced by Courtaulds.

Acrylic fibres are often blended with cotton or viscose rayon to create soft and light materials. They are often used for blankets, carpets, curtains and most furnishings. Knitwear is often 100 per cent acrylic fibre. In the early seventies 'high-bulking' acrylic fibres were developed for brightly coloured artificial furs and fleeces which could be machine-washed.

Carbon-Fibre Reinforced Plastics (CFRP)

If fibres made of polymerized acrylonitrile are passed through very high-temperature furnaces they become carbon fibres. These are exceptionally strong filaments, as fine as silk, which are used like glass fibres to reinforce resins for making strong light-weight structures such as racing car bodies (e.g. the Ford GT40) and compressor-blades for jet plane and hovercraft engines.

Carbon fibres were first made in Japan, but all carbon fibre technology nowadays results from research on the carbonizing of 'Courtelle' polyacrylonitrile fibre carried out by the Royal Aircraft Establishment at Farnborough and developed by the Atomic Energy Research Establishment. Rolls Royce has been both the largest manufacturer and consumer, while commercial carbon fibre is produced and exported by the Morgan Crucible Company ('Modmor') and Courtaulds ('Grafil').

UK: Diakon, Kallodent, Kallodentine, Kallodoc, Perspex, Transpex 1 (all ICI), Portex (Portland Plastics), Oroglas (Lennig Chemicals)
USA: Acryloid (Resinous Products and Chemical Co.), Barex 210 (Vistron Corporation), Jewelite (Pro-Phy-lac-tic Brush Co.), Lopac (Monsanto), Lucite (Du Pont), Plexiglas and Plexigum (Röhm and Haas Co.)
ITALY: Vedril and Vitredil (Montedison)
FRANCE: Altuglas
Trade-names for acrylic fibres include:
UK: Courtelle and Teklan (Courtaulds)
USA: Acrilan (Chemstrand), Orlon (Du Pont)
GERMANY: Douvel and Dralon (Bayer)
Trade-names for acrylic elastomers include:
UK: Miracle Bond (Avdel Ltd), Hanway acrylic caulk
USA: Cyanacryl (Cyanamid International), Hycar (B.F. Goodrich Chemical Co.), Thiacril (Thiokol Chemical Co.)

POLYETHYLENE (PE) *thermoplastic*

Everyone is familiar with polythene. The low-density grade is the most widely used of all plastics, and has achieved a phenomenal growth rate since its commercial introduction in Britain in 1942 and in America in 1943.

Polythene is the accepted abbreviation for the chemical name, polyethylene. This is the least complex of all the synthetic plastics, containing only carbon and hydrogen, linking on polymerization into a simple linear carbon-hydrogen chain:

Ethylene   Polymerized into   Polyethylene

It is usually recognized by its waxy feel. It is polymerized from the monomer ethylene, obtained from the cracking of petroleum or from natural gas, and is available in two main grades—low-density polythene (LDPE) and a tougher, high-density polythene (HDPE). As it is thermoplastic, polythene is easily moulded by the following standard processes: injection moulding (for most household goods including buckets, bowls, toys, novelties); extrusion (for sheets, films, tubes, plastic bags, cable insulation or in packaging); vacuum forming; blow moulding (for most squeeze bottles, milk bottles and liquid containers); and rotational moulding, which is like blow moulding but used for large containers such as drums and furniture. The disadvantages of polythene are that it has a low softening point (80°-130°C), it scratches easily, and it can suffer from stress cracking.

Ethylene can be polymerized by a high-pressure method and a low-pressure method. High-pressure polymerization produces low-density polythene. The molecular chain has many side branches which prevent the molecules from binding tightly to form a crystalline structure. As the molecules are loosely packed, they are low in density. Low-pressure polymerization produces high-density polythene. In this instance the molecular chain has few side branches, allowing the molecules to form a dense, linear, crystalline structure, producing a material which is harder, more rigid, more abrasion-resistant and with a higher softening point. But it is more brittle and not so translucent.

Like many other plastics, polythene was an accidental discovery. In 1933 chemists at ICI were carrying out experiments to see how chemical reactions altered when subjected to extremely high pressures. In one experiment a white, waxy substance appeared as a deposit in the apparatus, due to the presence of oxygen. At first this new substance could only be compared with the few existing plastics—PVC, polystyrene and acrylic—and by comparison it seemed a brittle material, without much future.

Eric Fawcett identified the new material and passed it on for evaluation. After that not very much happened until 1936 when the demands of the oncoming war increased the pace of research and the first British patent was filed. A pilot plant for testing high-pressure polythene was set up in England and in 1939, on the day that war was declared, production finally commenced, although full commercial production of ICI's 'Alkathene' did not start until 1942. It was to prove extremely valuable for tough, flexible insulation around radar and cable equipment. Production started in America in 1943.

In post-war England the end of the Utility period marked the beginning of the market in polythene consumer goods, and the demand for 'down-market' plastics bowls and buckets was so strong in America that by 1952 ICI was obliged to expand its manufacturing licenses. Polythene became so fashionable that by 1958 *House and Garden* magazine was advertising 'prestige' plastic flowers.

A different type of polythene resulted from researches by Professor Karl Ziegler at the Max Planck institute for Coal Research in Mülheim. ICI's high-pressure manufacture had been dogged by the ever-present danger of explosions. Professor Ziegler discovered in 1953 that it was possible to polymerize polythene at a much lower pressure, in fact at only slightly above normal atmospheric pressure, using special aluminium catalysts. This process produced a far more rigid polythene which is now called high-density polythene. The Ziegler catalyst could control the molecular structure of a polymer, and when Professor Giulio Natta in Milan polymerized polypropylene the following year, the era of made-to-measure synthetic materials had arrived. Simultaneously in America, two other companies had been working along similar lines and had developed their own low-pressure methods, namely the Phillips Petroleum process, and the Standard Oil of Indiana process. Nowadays either the Ziegler or the Phillips method is used.

Low-Density Polythene (LDPE)

Low-density polythene is flexible, odourless and non-toxic, making it suitable for many food and liquid containers, such as 'Tupperware' products. However, its widest use is as a film for plastic bags and carriers, for industrial packaging, and in

Tappeti (Nature Carpets) designed by Piero Gilardi. Sculpted from soft polyurethane foam and hand-painted. *Tappeti* began in 1966, hyper-realistic synthetic landscapes described as 'technological arcadia'. From early representations of wood fires, river beds and ladders lying in grass among fallen apples, the 'carpets' have grown larger and Gilardi has perfected his technique.

particular the shrink-wrapping of food. In the building industry it is increasingly used as a flooring underlay and damp-proof membrane.

In the early sixties the Bulgarian-born artist Christo developed an obsession with packaging into an art form, Packaging Art, by wrapping anything from nudes and motorcycles to whole coastlines and canyons in plastic film. His crudely wrapped parcels transform the contents into unresolved images.

High-Density Polythene (HDPE)

High-density polythene has good chemical and impact resistance and consequently is more suitable than LDPE for blow-moulded dustbins, bleach bottles, industrial barrels, and extruded sheet, pipe and monofilaments for ropes and netting. Its ability to withstand boiling and sterilization has made it invaluable in hospitals and schools where plastics are replacing many traditional materials. High-density polythene is ideal for children's furniture in that it is soft and resilient, non-toxic and easy to wash. Children can easily lift polythene chairs and build structures themselves, and they can withstand rough treatment without damage. Above all, it is noiseless.

The first structural use of polythene in furniture was a children's stacking chair for the municipality of Milan designed in 1964 for Kartell of Italy by Marco Zanuso and Richard Sapper. The chair was injection-moulded with separately moulded cylindrical legs, a feature now common in plastic chairs and tables. To give the box-shaped chair the flexibility necessary for it to be comfortable, the seat and back were striated, giving it a corrugated appearance and allowing air to flow round the body. The chairs are stacked by slotting the legs of one chair into the back of another.

In August 1969 Coca-Cola began to experiment with their first plastic bottle, blow-moulded in high-density polythene, moulded by Monsanto, and predicted as 'the forerunner of an avalanche of non-return bottles coming up in the 1970s' (DuBois, *Plastics History—USA*, 1972)—a statement that could only have spread forboding among those concerned about plastics waste disposal, and those addicted to the traditional glass bottle, originally designed in 1915 by Alexander Samuelson. The substantial weight of the glass bottle was replaced by the new plastic bottle which became lighter and lighter, from 57g in 1969 to 29g in 1973. In October 1969 the lettering was blow-moulded at the same time as the bottle, producing a similar effect to the traditional design, but the shape of the neck was changed. January 1971 brought the first resealable screw top. Finally, in February 1975 New York trembled as the rumour spread that the familiar Coke bottle was going to ultimately disappear. Monsanto were carrying out market tests with a completely new Lopac bottle, so-called because of its low oxygen permeability. It was described as 'the most

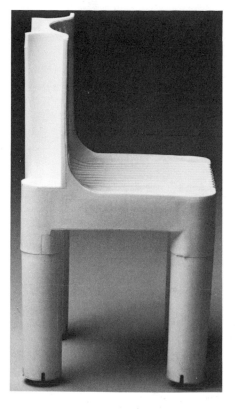

Childrens Stacking Chair 4999, designed in 1961 by Marco Zanuso and Richard Sapper for Kartell, Italy. Injection-moulded in high density polythene, it was Zanuso's first design in plastics.

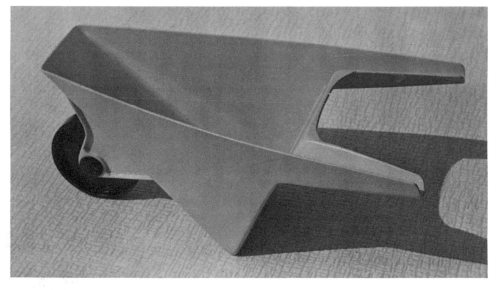

Wheelbarrow designed by Martyn Rowlands in 1960 for the Shell 'Design in Plastics' Competition, winning second prize. An early moulding in high-density polythene, the wheels are also plastic. It did not receive much publicity although the integral arms and the replacement of the traditional legs by large double-function hollow shapes have been copied since in commercial barrows. (Photo by courtesy of Martyn Rowlands)

The *Easy-Goer Cyclesafe* plastic Coca-Cola bottle moulded in Monsanto's Lopac acrylonitrile-styrene copolymer, which can be recycled. It was introduced for market testing in 1975. (The traditional Coca-Cola design has been replaced by a shape suitable for moulding in both plastics and glass.)

above right
Seven plastic Coca-Cola bottles illustrating design modifications carried out between August 1969 and January 1973. (The Coca-Cola Company)

thoroughly evaluated package Coca-Cola have ever marketed'. It was to be extremely light, impact-resistant, resealable and recyclable, and have a '38mm-wide mouth design for easier product pouring'.

However, the acrylonitrile-styrene copolymer was very expensive and major modifications were needed on production lines to manufacture the bottle in glass factories. Finally, during testing by the American Food and Drug Administration, the copolymer revealed a remote cancer risk, due to one of its additives, and Coca-Cola voluntarily withdrew the bottle.

Coca-Cola's chief competitor Pepsi-Cola had begun testing plastics bottles at the same time, using 'Dalar', Du Pont's PETP film, which was already in high-tonnage production. Large-scale testing had assured its success and PETP bottles for carbonated drinks will see a major growth in the future, along with other plastics.

Expanded Polythene Foam

If a blowing agent is added to the melt before moulding, it gives off a gas when heated which causes the polymer to expand into the honeycomb form known as cellular or expanded polythene foam. Only moulded in sheet form, early expanded polythene of the forties was fabricated into buoys and life-rafts and bonded to a variety of materials such as plywood or metal alloy in the construction of insulation panels for ships and aircraft, as well as refrigerators. Cross-linked expanded polythene foam is semi-rigid with a closed-cell structure and therefore has a low moisture absorption.

'Plastazote' has been made by BXL since around 1961, and is mainly used for packaging delicate instruments and equipment, and for making surgical splints. There are various ways of shaping it. As it does not flow as well as other plastics, only simple shapes are compression-moulded. A rough preform, about 10 per cent larger than the interior dimensions of the mould, is placed inside the mould and compressed into shape. However, it is normally moulded into sheets which are then fabricated to specific requirements. It can be sliced and then die-stamped with the shape of the object to be packaged, or it can be hot-hobbed, in which case the shape is melted into the foam. The sheets can be vacuum-formed on standard thermoplastic equipment or pre-cut into pieces, heated in an oven to 150°C and pressed around a cold shape.

Semi-rigid 'Plastazote' is the basis of an experimental instant splint.

106

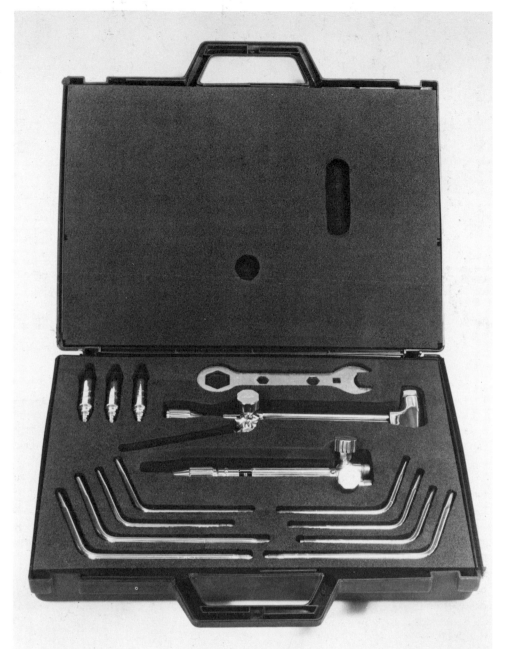

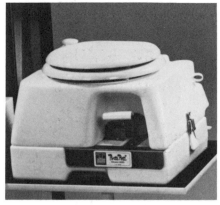

Porta-Potti 600 portable toilet, 1968, rotational-moulded in polythene by Thetford Engineering Corporation, Michegan. 60.9 × 60.9 × 45.7cm (24 × 24 × 18in.). The original model of a range which has been exhibited at the Museum of Contemporary Crafts, New York, and the Victoria and Albert Museum, London. A specialist design area in which plastics have revolutionized aesthetic appreciation. Its form is more like a picnic box with a built-in vacuum flask. (Thetford Engineering Corp.)

Portable oxy-acetylene cutting set protected in a shaped pack routed from laminated 'Plastazote' expanded polythene block, fitted inside a polypropylene case. (Customer Air Products Ltd. Photo Handford)

Bio-Degradable Polythene and Plastics Waste

'One day plastics will grow us over the heads if we are not watching to resolve their destruction'
(Translation of a statement by Hans von Klier in the Czechoslovak exhibition catalogue *Design and Plastics*, Museum of Decorative Arts, Prague 1972).

'Plastics is still the greatest growth industry of all time' (*Polymer Age*, September 1974). Consequently, it is increasingly important to plan for the future now and to solve the problems involved in the disposal of plastics waste. In 1973 a report on packaging published by the Friends of the Earth calculated that if all our milk bottles were made of plastic, the quantity of plastic tubing needed would be enough to girdle the earth every five or six days.

It is not yet certain whether, given time, plastics will totally degrade, whether they will rot in the end and return naturally into the ecological system. All that can be done at the present time is to look for ways of breaking down what has been created into a safe and useful form. Since it is the most common plastic used in packaging, research has centred around polythene. Being thermoplastic it can be recycled to a certain extent into new products and, like polystyrene, if incinerated it converts to the non-toxic gases carbon dioxide and hydrogen, natural elements in the atmosphere.

By 1972 plastics waste had become a regular target for criticism. A real fear was growing at the sight of the debris in the countryside and on our beaches. In 1970, out of eighteen million tons of refuse that was collected in Britain approximately 1·5 per cent consisted of non-degradable plastics. In Tokyo the figure was more like 8-10 per cent. The development of self-destroying plastics seemed the most logical solution to this problem.

In 1974 Pakex 74, the International Packaging Exhibition held at Olympia, London, took a defensive position on this issue: 'Waste is the new watchword, recycling the new remedy in the eyes and minds of those who, in some instances, regrettably are quick to condemn but slow to consider,' wrote the President of the Institute of Packaging in the foreword to the Pakex 74 catalogue. Fortunately, by the beginning of 1974 it had been established that polythene does decompose, although it is a lengthy process. Pioneering work carried out in Sweden found that additives mixed into the polymer reduce the decomposition time to one fifth. The additives make the polymer photochemically sensitive to heat and ultra-violet light, rendering it so dry and brittle that it eventually crumbles. This process takes the same length of time as the rotting of paper, or a pine needle, which needs eight years to disintegrate. At the same time chemists at Manchester University in England had been trying to convert waste plastics into food for people and animals. By mid-1974 they had succeeded in growing protein cultures and fungi on small amounts of oxidized polythene. In one sample they obtained 120 grammes of one type of fungus from 100 grammes of polythene. However, they could not guarantee it was harmless. Simultaneously an edible starch-based polythene was reported in Japan.

In mid-1975, after lengthy testing by George Griffin at Brunel University, England's first bio-degradable carrier bag reached the market, manufactured by Coloroll under the trade-name 'Byoplastic'. The polythene material incorporates particles of starch which, when buried, are attacked by an enzyme action in the soil, eventually to become harmless carbon dioxide and humus.

Other ideas have been put forward to utilize powdered degraded plastics, for example as a land-filler in 'soil mechanics'. But the best we can do, as the Friends of the Earth suggested in 1973, is to design packaging sensibly for re-use or to design specifically for re-cycling.

Ethylene-Vinyl Acetate (EVA) *thermoplastic*

The best-known of the polyolefin copolymers is EVA, ethylene-vinyl acetate copolymer, introduced in the mid-sixties. It is basically low-density polythene copolymerized with vinyl acetate. EVA copolymer contains more ethylene than vinyl acetate, so that it possesses the properties of polythene but with the additional rubbery flexibility of vinyl. Whereas low-density polythene and plasticized PVC can become brittle at low temperatures, EVA remains flexible, making it useful for refrigerator trays. It is also used for items such as WC pan connectors, inflatable toys, press-on bottle tops, handle grips, disposable baby-bottle teats, roofing underlay and all types of sheet and film such as for wrapping meat in supermarkets. EVA is moulded by the standard thermoplastic processes: injection moulding, extrusion, blow and rotational moulding.

UK: Alkathene and Alkathene EVA (ICI), Beetle (HDPE structural foam, BIP), Bexpand (structural foam, BXL), Carlona (HDPE) Shell Chemicals, Cobex (BX Plastics), Evazote,

Plastazote and Thermazote (BXL), Rigidex (HDPE, BP Chemicals), Sellotape (Sellotape Products, UK), Telcothene (Telcon Plastics), Visqueen (British Visqueen)
USA: Alathon (Du Pont), Fortiflex (Celanese Corporation), Marlex (Phillips Petroleum)
GERMANY: Baylon (LDPE and EVA, Bayer), Hostalen G (HDPE, Hoechst), Hostalen KD (LDPE, Hoechst), Lupolen (HDPE, BASF), Vestolen (HDPE, Huls)
ITALY: Eraclene (LDPE) and Eraclene HD (HDPE, ANIC), Moplen (Montedison)

POLYTETRAFLUOROETHYLENE (PTFE) *thermoplastic*

Polytetrafluoroethylene (PTFE) is very expensive to produce, and is therefore used only in small amounts. It is the most important of a group of plastics called the fluorocarbons, and is made from the monomer tetrafluoroethylene (TFE), a poisonous gas, with a molecular structure very like polyethylene. It has a similar waxy feel.

In the PTFE molecule the hydrogen atoms of the ethylene molecule ($CH_2 = CH_2$) have been replaced with fluorine atoms, giving tetrafluoroethylene ($CF_2 = CF_2$). The polymerization of tetrafluoroethylene is a dangerously explosive process, initiated by a peroxide catalyst in the presence of water.

Like polythene, PTFE is a thermoplastic, but it has the highest melting point of all thermoplastics, 330°C. It is therefore not very 'plastic', and does not flow easily when moulded, making it necessary to use expensive techniques when it is processed, which is usually by sintering.

PTFE was discovered accidentally in the late thirties by Dr R. J. Plunkett, a research scientist in the Du Pont Laboratories, and in 1941 a pilot plant began to produce it for military purposes, mainly as an electrical insulator. It was also used to protect the uranium in the atomic bomb. The first commercial production of 'Teflon', a Du Pont trade-name, began in 1948 at Parkersburg, West Virginia, but in Britain it did not appear until 1959, when it was produced by ICI under the trade-name 'Fluon'.

The replacement of hydrogen atoms in the ethylene molecule by fluorine gives PTFE an outstanding resistance to heat and friction, hence its use as an electrical insulator, for coating machinery parts and bearings, and for 'non-stick' food utensils. It is very pure and inert, and can therefore be used medically to make catheters or implants. Along with silicone it is a highly successful 'bio-polymer', and has been used in solid or textile form to rebuild bone and body tissue. It also has such an excellent resistance to chemicals that there is no known solvent for it at present. Just like nylon, PTFE has very low friction and is self-lubricating.

UK: Fluon (ICI)
USA: Fluorothene (Union Carbide Corporation), Teflon (Du Pont)

POLYAMIDE (PA; NYLON) *thermoplastic*

Nylon with a small 'n' is a generic name for a group of synthetic thermoplastics called polyamides, which are condensation polymers made from different types and combinations of dibasic acids and diamines derived from oil and natural gas. We are all familiar with nylon in fibre form as textiles and carpets, and in monofilaments as brushes and ropes, the most common forms of nylon; however, it is becoming increasingly used as a moulding material. In engineering it is being moulded into components to replace metal.

Moulded nylon is a very strong, tough, lightweight material. It has particularly outstanding abrasion resistance and anti-frictional properties, thus obviating the need for lubrication in gears, bearings, curtain tracks and zip fasteners, for example. It is also resistant to chemicals, and it has one of the highest softening temperatures of all the thermoplastics.

Since nylon is thermoplastic, it can be injection-moulded, extruded and rotational-moulded. Stock extrusions of rod, sheet or tube can then be machined on standard metal-working equipment into many other shapes for bearings, gears, adaptors, etc. Extrusions can be sawn, turned, tapped, ground, milled, drilled, punched and polished. Nylon 6, the usual grade for injection moulding, can also be cast in large standard sections or into mouldings which are too complex to injection-mould or to shape from stock extrusions. Glass-reinforced nylon can be chromium-plated or vacuum-metallized to simulate metal parts.

The chief disadvantage of nylon is that it is hygroscopic—it tends to absorb or lose water depending on the humidity in the atmosphere. A moulding can expand by as much as 3 per cent in maximum humidity causing changes in size and weakening the structure.

The first nylon to be produced, nylon 6.6, turned a new page in the history of plastics. It was the first deliberately created polymer and the first crystalline plastic material; i.e. it changed rapidly from solid to melt. It was also the first of the 'super polymers' of high molecular weight, with a repeating chain length of around 1,700 atoms. The repeating polyethylene molecule, written $(-CH_2-CH_2-)$, looks very simple alongside the complex nylon 6.6 molecule: $(-NH(CH_2)_6\ NHCO(CH_2)_4\ CO-)$. The unsaturated open bonds at each end are ready to join up through polymerization to make the chains that create plastics.

Following the discovery of nylon 6.6 a whole group of nylons with different properties was developed; these are the five main nylons in regular use:

For general use (stiff and strong):		Made from
Nylon 6	$-NHCO\ (CH_2)_5-$	Caprolactam
Nylon 6.6	$-NH\ (CH_2)_6\ NHCO\ (CH_2)_4\ CO-$	Hexamethylene diamine and adiptic acid
More flexible:		
Nylon 6.10	$-NH\ (CH_2)_6\ NHCO\ (CH_2)_8\ CO-$	Hexamethylene diamine and sebacic acid
Nylon 6.11	$-NHCO\ (CH_2)_{10}-$ (can be used as a coating and a film)	Omega amino undecanoic acid
Nylon 6.12	$-NH\ (CH_2)_{11}\ CO-$	Laurinlactam

The Russians have made a nylon 7 called 'Enanth' from amino enanthic acid.

Nylon 6.6 was synthesized in America in 1938. In 1927 an organic chemist at Harvard, Wallace Hume Carothers, had been commissioned to set up an open-ended research programme by the Du Pont Company. He started with Emil Fischer's discovery that proteins were basically long-chain structures, and that natural fibres like wool and silk were long-chain protein fibres. Carothers decided to study the chemistry of these natural materials and see if he could synthesize larger molecules in a predetermined pattern. The resulting material, nylon, was the first purely synthetic fibre. It was then developed by over two hundred chemists and engineers working together over ten years—a far cry from Baekeland's discovery of phenol-formaldehyde in his Yonkers laboratory.

In 1938 Du Pont's pilot plant produced the first commericial nylon, which was presented to the public in the form of 'Dr West's Miracle Tuft Toothbrushes'. Nylon replaced the traditional hog bristle, and most dentists agree that along with the new acrylic dental materials this was definite progress. Nylon toothbrushes are hygenic; compared with bristle they hold relatively little moisture and bacteria; they do not soften or fall apart so quickly and are more economical. For hairbrushes on the other hand, nylon builds up static electricity in the hair and tends to pull it out. Natural bristle distributes the hair's natural oil more effectively.

Nylon Fibres and Yarns

The alignment of the molecules in the structure of a polymer determines its suitability for making fibres and films. When thermoplastics are softened by heat the molecular chains of the material slide past one another. When cooled they remain in that shape until re-heated. To move these chains involves counteracting the powerful forces that hold the molecules together, known as the Van der Waals forces. The tighter the forces holding the molecules together the higher the melting point of the plastic material. Nylon has a crystalline structure, i.e. very tightly packed, making it the first artificial polymer suitable for mass-production of fibres.

Nylon fibres are made from nylons 6.6 and 6. The molten nylon is extruded through a steel spinneret (called melt-spinning), whereupon it cools and is wound onto bobbins. The filament is then hardened by cold drawing, i.e. it is stretched between two rollers turning at different speeds. Nylon fibres are made into monofilaments, multifilaments and staples. A monofilament is the extrusion of a single thread, like wire, which is cut up to make ropes, fishing line, gut for sports rackets, etc. Multifilaments are several fibres extruded through the die which are wound together to form one thread. Staples are shorter lengths of nylon fibre used for spinning textiles, often blended with other fibres. The cut lengths are from 3·5 to 20cm ($1\frac{1}{2}$-8in.).

By 1939 nylon yarn had mostly superseded natural silk in the manufacture of stockings and parachutes. Within the year sixty-four million pairs of nylons had been sold in America. The term used to indicate the thickness of nylon stockings, 'denier', describes the weight in grammes of 9000 metres of that particular nylon yarn. Monofilaments are high-denier, sheer stockings are low-denier. In the trade, the term denier has been replaced by 'decitex', the weight in grammes of 10,000 metres.

After World War II the potential of nylon yarn led to exciting changes in the manufacture of knitted fabrics. It could be dyed after bleaching and easily crimped, giving it the soft feel of wool. Today more than half our clothes are made of synthetic fibres—acrylic, acetate, nylon and polyester.

The development of heat-moulded nylon fibres in the fifties, together with 'Terylene' (polyester), contributed to the fashion for pleated skirts, a simple thermomoulding of the fabric, just as the 'perm' is the heat-moulding of hair. Stretch nylon jersey fabrics brought tremendous scope to the furniture industry to cover the new sculptural shapes which would otherwise have given upholsterers enormous headaches. Coated with polyurethane, nylon jersey introduced the 'wet look'.

Another minor revolution in the furnishing field brought carpets, once the luxury of the rich, within the reach of most people. Nylon 6.6 filament pile carpeting has outstanding endurance, although somewhat prickly, and is ideally suited to places with a lot of traffic, such as shops and stores. To make the carpets, filaments of nylon are electrostatically flocked onto an expanded PVC base material.

Synthetic turf was developed by Monsanto, USA, and first laid in the Astrodome at Houston (begun in 1960) where it set the pattern for many future sports pitches. Marketed as 'Astroturf', it is manufactured in three grades. 'Landscape', like a miniature version of the 'Pratone', is an open lattice of thick blades of polythene 'growing' at different angles. These are bonded to a mesh backing of synthetic fibre.

Astroturf synthetic turf, landscape grade, moulded polythene blades on synthetic woven backing. Developed by Monsanto, USA. Two other grades are made of nylon. (Photo Fritz Curzon)

It is a pleasant natural green tone, and its pigment is resistant to ultra-violet light. 'Action' is the grade used for sports grounds, in an unrealistic synthetic shade of green. It is made of 300 denier nylon ribbon fibre on a backing of woven black polypropylene, bonded into a synthetic black rubber underlay. The 'Patio' grade is the same as the 'Action' though fortunately a natural green, and bonded to a lighter rubber backing. The pile lies in one direction and feels like the rough hide of an animal. Other types of synthetic turf have followed, and are still being developed.

Glass-filled Nylon

In 1941 Du Pont introduced nylon moulding powder. The more recent development

Cado 290 Chair designed by Steen Østergaard, in 1970, manufactured by France and Son, Denmark. The first entire chair to be injection moulded in glass fibre reinforced nylon. Weatherproof and anti-static, it bolts together to form rows, and with a simple but ingenious stand, twenty-five can be stacked together (*right*). Foam cushions can be screwed to the seats. The minimal use of material illustrates its mechanical strength.

of nylon reinforced with glass, around 1970, has produced a material suitable for use in engineering. The glass filler or reinforcement increases nylon's resistance to wear and makes it much stronger and stiffer. It also lowers creep at higher temperatures and increases its electrical insulating properties.

Colourful glass-filled nylon now often replaces heavy die-cast metal in 'clam-shell' casings for electric power tools. Nylon is also suitable for moulding propellors of all sizes, from very large for boats and domestic fans to delicate inhalers such as the Fisons 'Spinhaler' designed by Martyn Rowlands for asthma sufferers. Glass-filled nylon is also incorporated into artificial limbs, such as wrist and elbow activators, as it does not irritate the skin.

UK: Beetle Nylon (BIP), Bri-Nylon (Nylon 6.6 fibre, Courtaulds), Enkalon (British Enka Fibre), Maranyl and Maranyl A 190 (glass-filled nylon, ICI)
USA: Antron and Cantrece (Du Pont), Caprolan (Allied Chemical Corporation), Plaskon Nylon (Adell Plastics), Zytel (Du Pont)
GERMANY: Durethan (Bayer), Perlon (Nylon 6, Bayer), Ultramid (BASF), Vestan (Bayer)
ITALY: Renyl C (Nylon 6, Montedison), Sniaform and Sniamid (Nylon 6, SNIA)
FRANCE: Rilsan (Nylon 11, Organico)

POLYURETHANES (PU) *thermoplastic and thermosetting*

Polyurethanes form one of the most valuable groups of modern synthetic plastics. The chemical structure of the polymer can be reorganized and modified into many forms: fibres, fabrics, elastomers and coatings, and flexible, rigid and structural foams. We wear polyurethane; we sit and sleep on it; we use it to build our houses, cars and machines; we coat caravans and factory floors with it, and seal windows, doors and stonework. And although it remains largely unknown and unnoticed, we have come to depend on it.

The formation of polyurethane is based on two fundamental reactions of isocyanates:

1) Isocyanate reacts with compounds containing hydroxyl groups to produce the urethane links in making the polymerization chain.

2) If water reacts with isocyanate, carbon dioxide gas and substituted ureas are produced.

In polymerization, di-isocyanates react with polyols (such as polyether and polyester) to produce the polyurethane plastics, but if this takes place in the presence of water and catalysts, with all the reactions co-ordinated, gas bubbles are liberated and trapped in the forming plastic. The result is a rigid or flexible open-cell polyurethane foam called expanded foam. Blowing agents can create the same foam but with a closed-cell structure, more useful for insulation. Additives determine the size of the gas bubbles and maintain the foam in an expanded state until the reactions are completed.

The two main types of foam are 'polyether' foam and 'polyester' foam.

In 1937 Dr Otto Bayer and associates in Germany discovered polyurethane polymerization, and during the war his laboratories developed the material, producing early rigid foams, polyurethane coatings, adhesives and fibres. In the late forties, the Monsanto Company and Du Pont began producing the 'one-shot' self-skinning foam process, and started commercial production of flexible and rigid polyurethane foam in around 1955.

Polyurethane Foam

Polyurethane foam, whether with 'open' or 'closed' cells, can be made flexible, semi-rigid or rigid, and in different grades or densities according to the strength required. Although it is not yet possible to make a fire-resistant grade, polyurethane foam can be given a fire-retardant additive which makes it self-extinguishing, a property which has become more critical as the use of foam increases, and especially as polyurethane emits toxic carbon monoxide and cyanide fumes when it burns.

opposite

Super Chair designed by Roger Tallon in 1969 as part of the 'Module 400' range, manufactured by Jacques Lacloche, Paris. Profile-cut polyether sheet glued to aluminium sheet with a pedestal base of cast aluminium. 49.5 × 59.4 × 125.7 cm (19½ × 23⅜ × 49½in.). More like a fakir's bed of nails, the dark grey foam is not usually associated with seating but packaging. (Photo Robert David)

Profile-cutting of 'Dunlopreme' polyurethane foam to produce an 'eggbox' effect. The sheet is distorted by spiked rollers as the cutter passes through the centre of the sheet. (Dunlop Ltd)

Polyurethane foam has two other important properties. Firstly, the density, that is the weight of the material per cubic foot, can be varied, and secondly, the foam can be made hard or soft. These properties are quite unrelated, in contrast to natural latex foam rubber where hardness or softness is related directly to density.

Polyurethane foam in block form is made by feeding liquid chemicals through a nozzle which moves backwards and forwards across a conveyor belt, or continuous trough. By the time the nozzle reaches the end of the trough, the mixture has expanded and cured (i.e. vulcanized) into a length of foam, known as a 'bun loaf'. Sheets and blocks, whether flexible or rigid, can then be machined to the required shapes and sizes on standard machinery. Flexible foam can be profile-cut, producing the textured 'egg box' sheeting used in packaging and adopted, for example by Roger Tallon for his chairs. Thin polyether foam sheeting, used as a backing for fabrics, is peeled from a revolving roll of flexible foam, just like veneer cut from a log. A thickness as low as 0.5mm can be sliced off in this operation.

To mould shapes from flexible or rigid polyurethane, the same chemical mixture is injected into a metal mould and either cured by applied heat, or cold-cured, the latter being a new process in which the foam cures itself. GRP moulds can be used for short production runs, although a smooth finish is not so easily obtained.

Flexible PU Foam

Flexible polyurethane foam is most familiar as a cushioning or insulating material, especially in furniture, automobiles, bedding and lagging for hot water pipes. Polyether is more resilient than polyester and therefore used more for cushions, whereas polyester is laminated onto fabrics and textiles. Cushions and mattresses are often built up with an inner core of denser polyester foam, with outer layers of softer polyether. All sorts of upholstery problems can be resolved with a mixture of densities and hardness. The acoustic and insulating properties of flexible foam make it a useful material for lining speaker and compressor cabinets.

It is also an excellent material for protecting delicate objects in transit, such as doctors' or photographers' equipment, with the foam contoured to the shape of the object. The surface is often flocked for 'luxury presentation'.

Being bacteriologically inert, polyurethane open-cell flexible foam is often used in hospitals, and is also an asset to asthma sufferers. Mattresses, and materials for geriatric treatment and for spinal supports can be made out of simple blocks. Polyurethane foam is odourless and can be sterilized. Applied to a burn as an instant dressing, it reduces the risk of infection and allows the burn to evaporate, even though it is in direct contact with the foam.

Flexible Foam Furniture

Flexible polyurethane foam liberated the furniture designer from many of the traditional constraints which had tied him to particular forms and structures. A constant flow of new plastic materials has prompted the creative imagination, and when the intrinsic properties of such materials are exploited, new shapes, new experiences, and new technologies quickly evolve.

Furniture designers, however, were slow to appreciate the future potential of polyurethane but used it first to develop and enhance existing forms. A chair was still unmistakably a chair, still 'domesticated' furniture. Olivier Mourgue's 'Djinn' chair is still traditional in the sense that it is obvious what it is, but it was the first design in furniture to move towards a much more sculptural and flowing form.

When rigid foam appeared in the late sixties, exactly the same sort of sculptural shapes that Mourgue and Paulin created expensively with metal tubing could be achieved but with half the work. Instead of an internal steel skeleton with webbing, diaphragms and reinforcement, the whole shell could be a one-shot cold-curing foam moulding, requiring only a surface layer of soft foam and a stretch cover.

In banishing the traditional steel frame the main problem is that of supporting the foam and keeping it in shape and protected. The 'Ciprea' chair (1967) by Afra and Tobia Scarpa, manufactured by Cassina, Milan, is like a traditional armchair in form, yet in structure it is quite different. It is made of flexible polyurethane foam, injection-moulded onto an ABS base inside the mould. The foam also has positions for buttoning moulded into it. It needed no more than upholstering in luxuriously soft leather, with castors fixed to the rigid base.

The next logical step towards all-plastic furniture was to find a way to incorporate the cover into the manufacturing process. Integral upholstery is still in the development stage. One method at the present time is either to dip the block of foam into PVC or to spray it on. There are disadvantages in this: when sat upon, the skin wrinkles, and after a time tends to separate from the flexible foam, and then breaks up. However, a sprayed PVC skin process has been patented which adheres to flexible foam, even when well crushed, without separating. Fabric flock is another spray finish, and is a softer surfacing material, but it tends to wear off in patches.

Once foam furniture became established, it encouraged the development of modular foam systems, now a standard feature of interior design. It is a convenient, practical and above all flexible way of buying and selling furniture, and reflects the move towards coordinated and 'environmental' interiors. Nanna and Jorgen Ditzel's foam blocks, Johannes Larsen's hinged shapes and Rupert Olivier's cube are variations on the theme. A wide variety of similar ideas has followed—now with

opposite above
Chaise longue from the *Djinn* series designed by Olivier Mourgue in 1964-65, manufactured by Airborne, France. Polyester foam over steel tube frame with a stretch nylon jersey cover and removable head rest. 60 × 73 × 65cm (23⅝ × 28¾ × 25⅝in.) high. With its foam wrapped round a sculptured steel skeleton, this chair initiated a new image in furniture.

opposite below
Early modular foam system, designed by Nanna and Jorgen Ditzel in 1952 but not manufactured until 1966. Geometric blocks of polyether foam covered in stretch fabric for building a variety of seats and supports. (Photo Louis Schnakenburg)

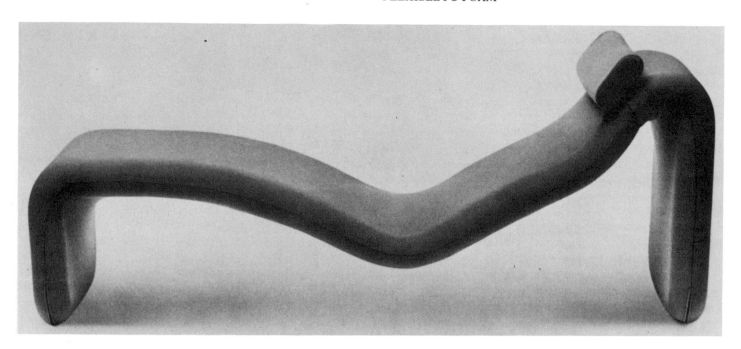

large-scale hinged beds and sofa units—while Italian designers have produced many geometric hinged blocks based on squares and circles, variously upholstered in everything from textiles to shocking pink sticky PVC. Many of these latter designs, however, are more exercises in manufacturing convenience- than in consumer comfort. From neatly packaged geometric solids designers have taken the concept a stage further into the realms of the articulated carpet. One example is Archizoom's 'Superonda' (Superwave) convertible sofa/bed for Poltronova, 1966. Floors became undulating waves of modular foam blocks, doing away entirely with traditional sitting positions, suitable only for the young and agile, if they could cope with the cost and the backache. Very few of these designs ever found their way into homes.

As space in the home becomes more precious, it is essential to consider space-saving devices more carefully. Ella Moody likens designers' fascination with modular units to playing with dominoes or with 'Lego'. Once the basic variants have been decided, it becomes a simple draughtsman's exercise to juggle with the plans of the units. But what the Italians did so well in the sixties was to expand simultaneously the horizons of the technician and the artist. Cini Boeri's 'Serpentone' needs no upholstery as it has a tough flexible skin integral with the polyurethane foam. The skin is formed as the foaming chemicals come into contact with the hot mould. However, it does not have the reassuring texture of a traditional fabric cover, and designers and manufacturers have searched for many years to find ways of achieving the ideal—the one-shot upholstered chair, a complete chair with its body and upholstery moulded in one operation. Attempts have been made by injecting both rigid and flexible foams into a mould already lined with fabric or sprayed with flock. It is an attractive ambition, which will one day supersede the laborious and expensive processes of bonding foam to wood or to plastic, of shaping cushions, and stapling tailored covers to frames. But there are problems involved, including loss of the washable cover, and the difficulties of repairing damaged fabric.

One method patented by Yorkshire Fibreglass Limited involves placing a GRP (FRP) chair shell into the female part of a two-part hinged mould. When the mould is closed, chemicals producing cold-curing flexible polyurethane foam are injected into the cavity left in the mould around the chair shell. It is simply removed and upholstered, fitted with a cushion and fixed to a variety of pedestals. Yorkshire Fibreglass has also experimented with fusing fabric to a chair shell with foam.

Serpentone (Giant Snake) designed by Cini Boeri, manufactured in 1971 by Arflex, Italy, with the collaboration of Vefer. The photograph shows the sections glued together and arranged round a room. The integral skin, industrial colour and visible mould parting lines illustrate a new Brutalism in furniture, in contrast to the civilized and aesthetically-worked designs of Joe Colombo. (Photo Masera)

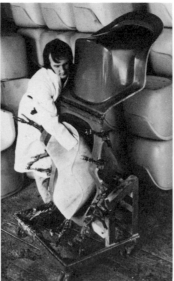

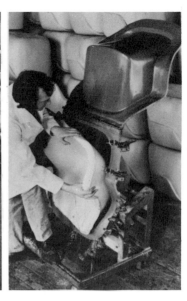

above
Sequence of photographs illustrating the 'Fibrefoam' process of injecting flexible cold-curing foam around a GRP chair shell. The shell is only ⅛in. (0.32cm) thick; the mould is also GRP. (Yorkshire Fibreglass Ltd)

The construction of *Serpentone*: removing a modular section from the mould. (Photo Bonetti)

Models *UP/5* and *UP/6* designed by Gaetano Pesce in 1969, manufactured by C & B Italia. Two of seven models from the *UP* range which includes a giant foot for use as a chaise longue, or 'pied long'. The chair and footstool of moulded flexible polyurethane foam are compressed into flat plastic envelopes for easy transport. The range was subtitled 'Release from Conventions' and was indeed a totally new concept in furniture, although further developments are yet to be seen. (C & B Italia)

The most innovatory design in furniture made from flexible polyurethane foam is the 'Up' series designed by Gaetano Pesce. The English word was chosen for these designs because the chairs literally spring up into their pre-moulded shapes before your eyes like an inflatable. Carrying a three-piece suite is reduced to the simplicity of carrying three flat boxes. The moulded foam shapes are compressed to one tenth their full volume in easily transportable airtight plastic envelopes. On cutting round the envelope the furniture is released and inrushing air expands it back to its original size and shape.

Up to this point the use of flexible foam in furniture remained within the accepted forms of that area of design. But freed from traditional technology, designers have also discovered that the potential of flexible foam was as unlimited as their imagination. If the Dadaists of 1915-22 had had access to these materials they would certainly have produced similar anti-conformist objects. The 'Mae West's Lips' sofa designed by Salvador Dali in 1936 was manufactured in polyurethane foam some thirty years later by the Italian company Gufram, and renamed 'Bocca' ('Mouth').

This glorious form of adulation spawned many descendants imitating various parts of the body—gaping mouth seats ringed with teeth, immense stuffed baseball-gloves, boxing gloves moulded in cold-curing polyether, and of course the giant foot by Gaetano Pesce. This realization of fantasy with foam stimulated designers into creating what is often known as 'fun foam furniture'.

Gufram, specialists in anti-design objects, have also manufactured Piero Gilardi's 'Sassi' ('Pebbles'), various sizes of boulders made from 'Guflex' expanded polyurethane foam and painted realistically with 'Guflac' sparkling lacquer. When placed casually on a floor they look very convincing, but they are soft not hard. The lacquer finish is both waterproof and weather-resistant, so that theoretically they could be deposited in the garden like the real thing. Gilardi also designed a version large enough to sit on called 'Sedil-Sasso' ('Stone-Seat'), and pebbled tiles 50cm × 50cm (20in.) called 'Pavépiuma', to be laid down with double-sided tape.

left
Sassi (Pebbles) and *Sedil-Sasso* (Stone-Seat) designed by Piero Gilardi in 1967, manufactured by Gufram, Italy. Moulded in 'Guflex' polyurethane foam painted with waterproof 'Guflac' lacquer. (Photo Sartor)

below left
Pratone (Large Meadow) seating unit designed by architects Ceretti, Derossi and Rosso (Gruppo Strum) in 1970, manufactured by Gufram, Italy. Cold-foamed 'Guflex' polyurethane painted with 'Guflac', 110 × 120 × 82cm (43¼ × 47¼ × 32½in.). A grossly enlarged sod of turf, almost identical to the 'Astroturf' (see page 111), and more likely to devour you than to invite you to lie in it.

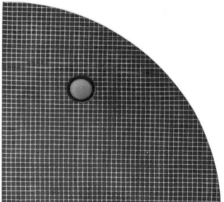

above
Torneraj (You'll Come Back) designed by Ceretti, Derossi and Rosso in 1969, manufactured by Gufram, Italy. Cold-foamed 'Guflex' polyurethane painted with 'Guflac'. A baffling quadrant shape to the uninitiated which in fact is not solid in section; the seating area drops into a hollow centre. Polyurethane foam has made possible furniture of a much more abstract character and the design of seating has extended to formats which appear to negate their function. 90 × 90 × 90cm (35⅜ × 35⅜ × 35⅜in.).

Rigid Polyurethane Foam

The chemical mixture for producing rigid expanded foam is basically the same as that for flexible foam, with a blowing agent and catalyst added to the polymer compound to produce the gas that causes the expansion. In this instance, however, the two liquids react together quickly and the heat generated causes the polymer to vulcanize and set in a cross-linked irreversible structure.

Rigid polyurethane foam looks similar to PVC and urea-formaldehyde foams, and is used in much the same way, for example in packaging and insulation, but is more expensive than polystyrene. Rigid polyurethane foam has in many instances replaced cumbersome and expensive wooden crates. An object can be simply wrapped and placed in a cardboard box and two-part resin injected around it, which foams and sets in only a few seconds into either rigid or semi-rigid foam according to the mixture. The result is 80 per cent lighter than a wooden crate.

For insulation purposes a cheaper, low-density grade is used, and aerosol packs are available for sealing leaks and cracks, or for dampening sound.

Rigid polyurethane foam has been used to construct low-cost buildings, particularly for instant shelters in the aftermath of natural disasters. When an earthquake in Peru killed 80,000 people in May 1970 the German firm Duttger provided polyurethane domes by spraying the foam onto inflated balloons. The balloons rotated on a disc until a wall thickness of 10cm (4in.) had been built up, a process which took one hour, and then the balloon former was deflated. Next, plastic windows and doors made of PVC blinds were fitted, and the 'Igloniums', as they were called, were finally sprayed with a protective coating of latex rubber. They are now five years old, and many are no longer temporary waterproof homes but have become permanent examples of survival and adaptation.

When the earthquake of autumn 1975 made 2,000 villagers homeless in Lice, Turkey, Oxfam responded with similar instant shelters built from a mobile lorry containing the necessary chemicals and spray equipment (including a generator), surmounted by a pulley system on top. It was so compact that a lorry and enough chemicals to build 800 homes could be fitted into two aircraft. The foam was sprayed by men in protective clothing onto the waxed interior walls of a hexagonal aluminium mould, which looked very much like a cake tin. Elements for windows and doors were fixed in position on the mould before spraying, and the wall was built up to a thickness of ten centimetres. Once the expanded resin had set, the mould was hoisted up on the pulley and mounted on top of the lorry, like a bucket revealing a sandcastle.

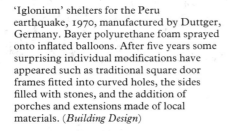

'Iglonium' shelters for the Peru earthquake, 1970, manufactured by Duttger, Germany. Bayer polyurethane foam sprayed onto inflated balloons. After five years some surprising individual modifications have appeared such as traditional square door frames fitted into curved holes, the sides filled with stones, and the addition of porches and extensions made of local materials. (*Building Design*)

Oxfam's mobile 'factory' at work moulding rigid polyurethane shelters after the 1975 earthquake at Lice, Turkey. (*Building Design*)

This approach was somewhat cheaper than Bayer's and was also faster, moulding a complete home in twenty minutes. As it was made from the inside the work could be carried out in any weather conditions.

The avant-garde French sculptor César (Cesare Baldaccini) has used rigid polyurethane foam dramatically to exploit its chemical properties. At the Salon de Mai in the Musée d'Art Moderne in Paris in 1967 he exhibited a 'Big Orange Expansion', characterized as an 'enormous orange vomit' by Grégoire Muller in *Art and Artists*; it was made with forty litres of chemicals and extended for more than five metres along the floor. In a similar 'Expansion' at the Tate Gallery, London, in 1968, César poured out three batches of polyurethane chemicals onto a polythene-protected floor, each lot weighing sixty pounds. Within a couple of minutes the resin had expanded and set. By the end of the evening these gigantic chemical shapes had been ripped apart by souvenir hunters in evening dress, and pieces were even signed by the artist.

Recently César has shown more concern with the durability of his sculptures. He began to reproduce the foamed shapes in durable materials like wood, bronze, brass and stone; and later he found a way of preserving the originals beneath a GRP skin, rubbed down and sprayed with a synthetic lacquer. These preserved originals could be used as formers for making the moulds for mass-production in metals or other materials. He even copied one original gush of polyurethane with marble, the final gesture against impermanent Process Art, resulting in a strange paradox—stone simulating plastics.

Expansion by César. Metallic grey GRP skin over rigid expanded polyurethane foam. 213.4cm (7ft) high. (Photo André Morain)

below
Expansion no 42 by César, 1973. Pink mother-of-pearl GRP skin over rigid expanded polyurethane foam. A rush of foam carries with it several faces of the same man. (Photo André Morain)

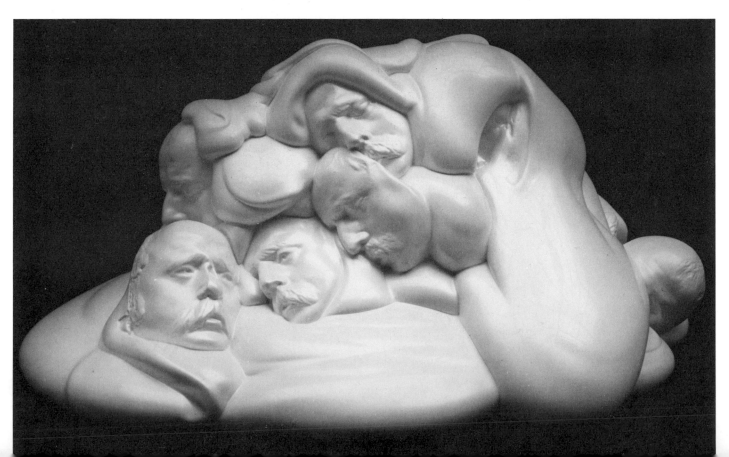

Rigid-Foam Furniture Shells

Like expanded polystyrene, rigid polyurethane can be a complete structural material in itself, and, when used in the correct shapes, can make self-supporting structures for chairs and settees. Unlike polystyrene, reinforcement can be moulded into the shape, and it is a cold-curing foam, needing no heat or pressure. It can therefore be formed in cheap moulds, which cuts down tooling costs considerably.

In 1966 the British company Hille formed HIA Plastics to exploit the new rigid foam and to produce short runs of rigid polyurethane shells in different designs moulded in cheap GRP moulds. The first chair moulded in this way was 'Cumulus' by the Hille designer Robin Day. Although it looks simple there are strict limitations in designing chair shells. The material must be of the correct thickness, although this

Chair shells of rigid expanded one-shot polyurethane showing the variations possible from one basic mould. (ICI)

can be controlled to a certain extent by increasing the density in thinner sections, or moulding in reinforcements. Methods of covering and upholstering the shells are as important as correct moulding. As two designers commissioned by Hille, Fred Scott and Peter Murdoch, wrote: 'The designer is now in the position of a sculptor faced with a solid block of stone. He must use his new freedom with due regard to form, comfort, cost and marketability.'

Foams can also be built up without a mould or former to produce organic shapes, in the same way as some birds make their nests. Gunnar Aagaard Andersen has constructed furniture by using the basic two-part chemical mixture as used above, but with much less sophisticated equipment. He poured bucketfuls of resin by hand, one after the other as each layer set, to make a chair at present housed in the Museum of Modern Art, New York. It is a 'curious brown anti-object which can still be called a chair', according to Arthur Dexler, Director of the Department of Architecture and Design at the Museum. Using the natural brown colour of polyurethane foam and looking like a chocolate cake dripping with icing, it has aroused a lot of controversy, although its most severe critics admit that it is extremely comfortable.

Armchair, Gunnar Aagaard Andersen, 1964. Rigid polyurethane foam 76.2cm (30in.) high made at Dansk Polyether Industri, Denmark. (Museum of Modern Art, New York)

Rigid PU Integral-Skin Foam
Polyurethane Structural Foam (PU SF)

The development in Germany in the late sixties of integral-skin polyurethane foam increased its potential versatility. Rigid integral-skin foams are often referred to as polyurethane structural foam.

The term 'integral-skin' refers to a flexible or rigid foam (such as used by Boeri for her 'Serpentone') which has a tough surface skin formed during the moulding process so that no extra finishing should be necessary. The skin is formed when the expanding cells of the foam come into contact with the moderately heated surface of the mould and compact. The moulds for one-shot polyurethane moulding can be cheap because of the low pressures involved, and details or patterns etched into the surface of the mould can be accurately reproduced — for example, moulds cast from timber originals give a convincing illusion of its grained surface.

The 'In Chair 301' designed by the Italian architect Angelo Mangiarotti for Zanotta in 1969 could only be made successfully in integral-skin rigid foam. The surface is so smooth that the seat and back are striated to prevent the sitter from

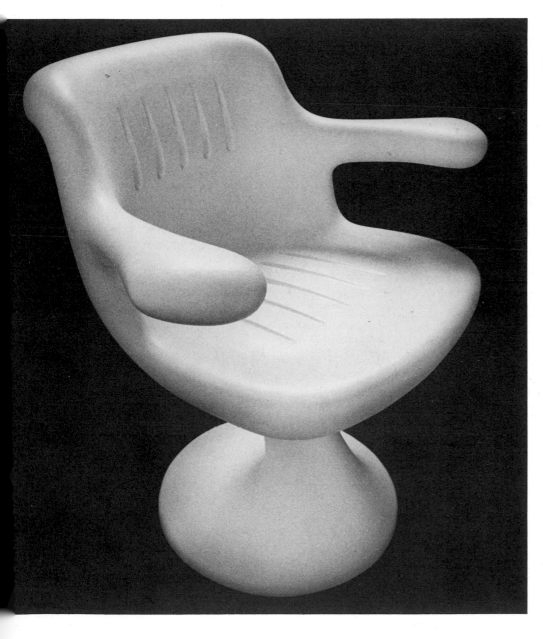

In Chair 301 office chair, designed by Angelo Mangiarotti in 1969, manufactured by Zanotta, Milan. Integral-skin rigid polyurethane foam. 68 × 57 × 72cm (26¾ × 22½ × 28⅜in.). (Photo Giorgio Savini)

127

Kangeroo chair No. 2003 (armchair) designed by Ernst Moeckl in 1971, manufactured by Horn KG Rudersberg. Injection-moulded in Bayer's rigid polyurethane structural foam. $58 \times 57 \times 76.5$ cm ($22\frac{7}{8} \times 22\frac{1}{2} \times 30\frac{1}{8}$). Over a cellular core, the skin is 1-2mm thick, sprayed with polyurethane primer and polyurethane colour. The chair illustrates the changing appearance of plastics furniture. As the stuffed chair was reduced to a metal cantilever, the generous forms of the 1960s are giving way to feats of engineering now possible with structural foam. Right angles can be moulded in structural foam, impossible with most other plastics.

slipping. Mangiarotti experimented with many shapes and produced a range of designs to fit the same base: stools with high and low backs or no backs at all. Although the 'In Chair' is reminiscent of a barber's chair, its globular appearance reflects an imaginative and somewhat humorous approach to design.

In the early seventies structural polyurethane foams moulded by a modified injection-moulding process were developed mainly in Germany with Bayer's 'Baydur' resins. This process produces mouldings which are much thinner in section, use less material, yet are just as strong. The material still has a foamed core with a solid outer skin, and wall thickness can vary from 4mm to 50mm. It has been used to mould TV cabinets, shelving units, tables and chairs. A child's cot has been made with each entire side as one moulding. A similar cot in timber would have required fifty-six jointed components. Parts for the Hammond organ have been moulded in polyurethane structural foam, where liquid chemicals have simply been poured into silicone rubber moulds to expand and set. This kind of low-pressure process leaves the characteristic swirling pattern on the surface of the moulding, but is not as cheap as polystyrene structural foam.

Polyurethane Coatings

Polyurethane resins make the familiar tough, hard, yet flexible paints, and are also used extensively to coat card and paper. Polyurethane coatings have been marketed in Germany since around 1968. Like the polyurethane foams, they are based on a two-component system, that is, a basic resin with a curing agent added to initiate the setting reaction.

Polyurethane coatings are non-abrasive and oil-resistant and can therefore be sprayed onto concrete factory floors. They resist chemicals and acids, and have been used to line wine presses. The glossy white caravans that abound on European caravan sites are sprayed with ultra-violet-resistant polyurethane that was developed after it was found that polyurethane white paint turned yellow.

The advent of polyurethane and PVC 'wet look' fabrics around 1970 was the beginning of a vogue for coated nylon fabrics. But just like the accidents that happened to plastics in the forties, the side-effects with 'wet look' fabrics showed that plastics are still subject to experiment and that by no means all the answers have been found. In 1974 polyurethane 'wet look' upholstery began to fall apart on furniture only two or three years old. 'Research experts are unable to predict beyond eighteen months how the fashionable polyurethane-coated fabrics will wear,' reported the London *Times*, 'but this has not stopped manufacturers and retailers selling such suites at £300 or more with no warning to the public.'

Polyurethane Elastomers

Like many other plastics, polyurethane can be produced in a synthetic rubber form. Polyurethane elastomers are very strong, highly resistant to oils, and wear well, and can therefore be found in industrial applications, cast into solid tyres, bearings, bushes and gaskets, shoe soles and heels, automobile parts and moulds for one-shot polyurethane moulding—in fact, in any situation where it is a suitable replacement for rubber. It is manufactured in many grades, from rigid to liquid.   Polyurethane moulded 'wedge' shoes, identifiable from their dark brown and black colours, have achieved enormous popularity. Many of the shoes are made entirely from polyurethane with flexible plastic uppers and elastomeric soles.

In fibre form, elastomeric polyurethane is not as well known as it deserves to be. The first polyurethane fibre was discovered during World War II by the Germans, who named it 'Perlon U'. Much later, in 1958, the first polyurethane rubber-like fibre was produced in America by the Fristene Company. Elastic fibres like these are now generically known as 'Spandex' fibres, of which the best-known is 'Lycra' by Du Pont. They are not generally used on their own but spun in combination with stronger fibres like rayon and nylon. Synthetic elastomeric fibres like these have gradually replaced rubber in the manufacture of support garments, stockings and underwear, with a subsequent increase in the occurrence of rashes on sensitive, allergic skins.

USA: Adiprene (Du Pont), Corfam (urethane reinforced with polyester, Du Pont), Estane (Goodrich Chemical Co.), Genthane (General Tire and Rubber Co.), Lycra (elastomeric fibre, Du Pont), Spandex (generic name for polyurethane elastomeric fibres by Fristene Co.) GERMANY: Baydur (HD rigid integral-skin foam, Bayer), Bayflex (semi-rigid integral-skin foam, Bayer), Desmodur and Desmophen (coatings, Bayer), Vulkollan (elastomer, Bayer) ITALY: Glendion/Tedimon (structural foam, Montedison)

POLYESTER *thermoplastic and thermosetting*

Polyester resins are a varied group of condensation polymers which result from the reaction of a dibasic acid with a dihydric alcohol, of which either or both can be saturated or unsaturated compounds. All the raw materials for making polyester resins are derived from petroleum.

If the resins are saturated, there are no free radicals (links) in the molecular chain for cross-linking to take place, therefore the material remains thermoplastic. Saturated polyesters (SP) can be spun into fibres, or extruded as sheet and film.

If the resins are unsaturated they have free radicals in the chain to which cross-links can be attached, and can therefore be co-polymerized into cross-linked thermosets. Unsaturated polyester resins (UP) are cross-linked by an unsaturated monomer, styrene being the most commonly used, or diallyl phthalate, which produces a more flexible resin than with styrene. The properties of the polyester resin depend on the monomer used. Diallyl phthalate is also used as a plasticizer in other plastics, rendering them more adhesive and flexible. Though expensive, UP is also used for embedding, casting, and encapsulating electrical equipment; it is very hard and as clear as glass.

Clear unsaturated polyester resins are used for embedding zoological and natural specimens, and a light-stabilized resin is available which is resistant to yellowing.

Of the two types of resin, saturated and unsaturated, the unsaturated resins are more common since they are used for making GRP mouldings. The basic polymerization reaction of an unsaturated polyester resin requires that either the acid or the alcohol is unsaturated. They react together with the cross-linking agent as catalyst or

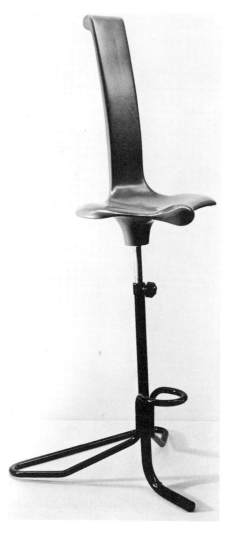

L'Appoggio (The Back Support) designed by Claudio Salocchi in 1970, manufactured by Sormani, Como. Injection-moulded 'Baydur' polyurethane structural foam on adjustable metal frame. Based on the principle of the monks misericord, *L'Appoggio* is very practical in situations where people have to stand yet need to take the weight off their feet. Salocchi once said, 'I am frightened by the image of a world made of plastics inhabited by men made of plastics.' Ironically a group of these chairs look just like a crowd of plastics men.

hardener, and are heated either externally or by an added accelerator. The chief properties of unsaturated polyester are electrical insulation, heat insulation, and resistance to chemicals and weather. Unreinforced unsaturated polyester resins, also called alkyds, are used as protective coatings. Many ordinary hard-gloss paints are alkyd-based, such as Du Pont's 'Dulux' gloss. The alkyd imparts a degree of flexibility to the paint, unlike other more brittle oil-based paints, such as polyurethane.

Alkyd resins were first documented around 1901 in England by W.J. Smith. When the General Electric Company in America brought out their first alkyd resin in 1926 they named it 'Glyptal', after its components glycerol and phthalic anhydride. A thermosetting condensation resin, it was first used as an electric insulator, bonded with mica flakes, and then commercially as an adhesive coating. This development was a stroke of luck for the nascent car industry, which in those days hand-finished each car with a laborious series of processes—sanding, priming, sanding again, lacquering, etc. The new resin was simply sprayed onto the primer and cured by passing the car body through an oven to produce a weather-resistant high gloss. In the forties alkyds were widely used for stove enamelling and mixing paints. Small compression mouldings were also made from this resin, exploiting its electrical and chemical resistance to mould handles and knobs for electrical applicances, plugs, sockets, and housings for electrical hand tools.

In 1941 casting resins were developed as a substitute for glass, and in 1942 low-pressure laminating resins were introduced and the first radomes (spherical radar-proof shelters) were laminated in the United States and Britain. Whereas other laminating processes, for example phenolic and urea-formaldehyde laminates, require high pressures and high temperatures, the advantage of polyester resin is that the resin can cure at room temperature, or more quickly in warm ovens, with little or no pressure. Polyester resin is thus cheap to process, although moulding techniques are often labour-intensive.

The earliest GRP mouldings were made during the War, such as the first GRP boat hull in 1942, and radomes for tracking transmitters carried in balloons. Cold-setting unsaturated polyester resins appeared in 1947-8. They benefited one little-known area of plastics application, that of erosion casting. This is the process for making casts of parts of the human body, such as the nervous system and arteries, used by professors of medicine to teach their students. The casts are made by filling a particular part of the body with resin, and eroding away the surrounding tissue, leaving a brightly coloured three-dimensional sculpture.

The chosen part—lung, brain, or heart—is injected with highly coloured polyester resin with a catalyst, which flows along the veins or arteries. This is clearly very delicate work and is often performed underwater so that the tissues retain their full three-dimensional form. Water also helps to slow down the curing process by keeping the temperature low, allowing time for the intricate operation. When the resin has hardened the surrounding tissue is eroded away in acid, thus revealing the solid coloured cast, like a bright coral. The casting can be pruned like a plant, so that the moulding is selective. The final cast is sprayed with a transparent protective coating of polyester resin.

Glass-Fibre-Reinforced Plastics (GRP; FRP)

'Made of fibreglass' (or glass fibre) is a common term although it is inaccurate, like the use of 'Bakelite' to describe all types of phenolic resins, or 'Perspex' and 'Plexiglas' for acrylic. 'Fibreglass' is in fact the trade-name of the glass-fibre products made by Fibreglass Limited. The correct term is glass-reinforced plastic (GRP) in the UK and fibre-reinforced plastic (FRP) in the USA. GRP mouldings are one of a large group of fibre-reinforced matrix composites. Although glass is the main ingredient, the fibres vary widely from soya bean and carbon, to jute, sisal, asbestos and boron. Like many other thermosetting resins, the polyester has no strength on its own, but with the addition of reinforcing fibres it becomes a very tensile material

opposite
Oranges and Lemons, Michael Sandle. 1966. GRP, brass, 'Perspex' polymethyl lacquer. 182.9 × 487.7 × 274.3cm (6 × 16 × 9ft). (Tate Gallery. Photo by courtesy of the artist)

with high impact strength and a good strength-to-weight ratio.

A shiny surface and a rough underside, with the reinforcing fibres clearly visible in the resin, make many fibre-reinforced mouldings easily recognizable—translucent resin used for corrugated roofing sheets actually shows the fibres. But is not so easily recognized when pressed in a matched mould, when both sides of the pressing have a glossy surface finish.

Although strong, GRP distorts badly under compression, although it rarely cracks or breaks. Until recently it was also combustible; the three-wheeler GRP invalid cars in Britain were known to burn up. For architectural use, GRP is available with class 0 fire rating (i.e. non-combustible throughout), or classes 1 and 2 where the requirement is only for control of the surface spread of the flame.

GRP is enormously versatile with a very wide range of applications, including toys, vaulting poles, sports rackets, children's slides, switchgear and cable ducts, guttering, entire houses and outdoor applications from aerospace to automotive and marine. British Rail's new high-speed train is moulded in GRP. Until 1975 the world's largest GRP-hulled ship was *HMS Wilton*, launched in 1972, a 2,000 ton non-magnetic minesweeper moulded in BP Chemicals' 'Cellobond' resin. GRP underwater maintenance units operated by frogmen clean the undersides of tankers.

Nearly all these products are factory-made, but fibre-reinforced plastic is one of the few plastic materials that have proved practicable for amateur use, whether for customized motor-cycle cowlings, car-repairs, garden gnomes or sculptures. There have been some amazingly inventive uses: 'Mr W.M. of K. had a cracked hatching turkey egg. He repaired it with "Isopon" and put it into the incubator. Four weeks later the egg yielded a baby turkey.' 'Mymryn, an eighteen-month-old sheepdog, from Cefn du Mawr, Anglesey, had to have one of her hind legs amputated . . . . The owner contracted Mr Robert Owen, a well-known local engineer and inventor. He built Mymryn an artificial leg entirely from "Isopon", and the dog has now been working cattle and sheep for two years.' ('Isopon' catalogue, W. David & Sons Ltd, London.) The excellent weathering properties of GRP make it a tempting alternative to pressed steel for cars, to replace those parts which disintegrate through rust, although little has as yet been used commercially in this way. The Chevrolet Corvette of 1955 was the first production model with a moulded FRP body, but was one of only a few FRP cars made for many years. Now the use of GRP is mainly confined to small parts and limited runs—the nose of the John Player Special racing Lotus, dune buggies, the Bond motor cars, grilles, seats and head-rests.

Chevrolet Corvette, 1955, the first production car with an FRP body. A traditional two-seater sports car, hand-moulded in glass fibre reinforced polyester. 'Despite a lamentable lapse into camp styling' (*Industrial Design*), it was one of the only FRP cars available for many years. (General Motors Corporation)

GRP (FRP) Moulding Materials

Glass-fibre reinforcement is available in a wide variety of forms, designed for different moulding processes. Of the three kinds of matting, chopped strand mat is made of short lengths of glass filament bound together with resin; continuous filament mat is made with longer filaments and less resin, suitable for laying over curved and complex shapes; and surface tissue, a finer matting, is used as a smooth top surface over coarser matting below.

Glass rovings are used in spray guns. This is glass in thread form, made from glass filaments wound onto large bobbins. The spray gun chops the yarn and mixes it with resin, catalyst and activator before the mixture is sprayed.

For mouldings requiring greater strength, either woven glass cloth or woven rovings are used, looking rather like rush matting. All the preceding types of glass fibre shimmer like silk—an illusion, for glass fibre is very irritating and the resins burn the skin.

The resin itself is a thick liquid, usually a combination of an unsaturated polyester resin with an unsaturated monomer, such as styrene, together with a catalyst or curing agent, which initiates the cross-linking, and an activator (or accelerator) which speeds up the reaction.

Two other types of polyester moulding materials have appeared more recently, both pre-mixed compounds with good chemical and electrical properties. SMC (sheet moulding compound) is a continuous strip of chopped rovings coated with resin (the pre-mix), formed between two sheets of polythene on a moving belt. When the polythene is stripped off, the pressed, pastry-like slab is cut to required lengths and moulded in hot compression or transfer moulds (hot-pressed). This method can produce panel mouldings up to 35 per cent cheaper than hand-lamination and with patterned or textured paper incorporated into the top layer.

DMC (dough moulding compound) or BMC (bulk moulding compound) is basically the same pre-mix as SMC but is in the form of a dough. Cut to size, shallow mouldings can be compression-, transfer- or injection-moulded by the newly developed techniques for thermosets and are used for moulded window frames, tabletops and road signs. Where deeper mouldings are required, preforms are made first on perforated metal formers, which are roughly the same shape as the final moulding. They are cured and then placed in the press. Fire-retardants are sometimes added to both DMC and SMC for electrical use.

GRP (FRP) Moulding Processes

Contact or hand lay-up is the simplest process, needing no pressure or heat, and is used for short runs. Moulds can be made from the cheapest materials—wood, plaster, and even GRP itself—but it is the most labour-intensive process. One side of the moulding will always be rough, while the other side, laid next to the mould, whether a male or female mould, will be shiny and smooth. Matting is simply pressed against the mould with resin on a brush or roller, and left to dry at room temperature, although the moulding will cure within minutes if passed through an oven of up to 100°C. There is no limit to the size of the moulding.

Wet spray-up is a speedier, mechanical process which makes use of the same cheap hand lay-up moulds. The glass strands are chopped in a gun which sprays the two components directly onto the mould where they mix and gel.

For very large or intricate mouldings a rubber-bag (or vacuum/pressure-bag) can be used, which is vacuum-sucked over resin-impregnated matting laid over a metal former. An overall, even, atmospheric pressure compacts the laminate against the former. This process is similar to laminated plywood forming, but with less pressure and no heat.

Cold and hot matched-metal moulds are used for higher production runs where a good surface is required on both sides. A matched-metal mould consists of two parts, male and female, that fit into each other. The male die closes on the female mould and squeezes out excess fibre and resin (the flash), which needs to be removed afterwards.

In cold press moulding little pressure is needed. The resin is cured in the mould with self-generated heat. This is an economical process for instances when the hand lay-up technique is too slow. Cold press mouldings can also be carried out with glass fibre impregnated with phenolics and epoxides.

Hot press mouldings are similar but are produced at a faster rate since the mould is heated and higher pressures are induced by a hydraulic press. Up to fifty mouldings per hour are possible from one mould. Higher heat and pressure demand moulds of tooled steel, chrome-plated for the best results, or of aluminium alloy or cast iron. It is an altogether more expensive process but justified in terms of mass-production.

For matched-metal moulding the glass reinforcement, whether in the form of matting, preforms, DMC or pre-cut lengths of SMC, plus resin and chemicals, is placed in the mould, and after heating and curing the moulding is ejected.

The process of filament winding was developed around 1946 and produces hollow tubes and spheres capable of withstanding extreme pressures, such as pipes and containers, and parts for space vehicles. Resin-impregnated glass filaments or rovings are wound onto a revolving mandrel, which is later removed.

In centrifugal casting the inside of a mandrel and centrifugal force are used to mould glass fibre and resin into cylinders. A recent development is the injection moulding of cold DMC into a hot mould. The dough flows well and forms smooth, glossy mouldings. Polyester injection mouldings, reinforced or un-reinforced, are a good substitute for metal as they are rust-proof, self-lubricating and have a high fatigue endurance.

Pultrusion is one of the most recent techniques for moulding GRP, and is already well-established for moulding carbon fibres. It is the reverse process to extrusion. Glass-fibre matting or rovings, or a mixture of both, is pulled with great force through a long steel die. Inside, the reinforcement is impregnated with resin and then heated as it moves through, to emerge cured and solid. A pultrusion has great linear strength, since the glass filaments are kept very straight and in line. It is used for moulding continuous profiles such as rods, tubes and sections, similar to extrusions, and can produce rods with a central core in one operation. Pultrusion mouldings have excellent electrical and thermal insulation properties as well as being non-corrosive, and are used for railway and electrical ducting, fuse holders, ladders and handrails, gutters, flag poles, skateboards and bows and arrows.

GRP (FRP) Furniture

The development of GRP can be traced most comprehensively by studying its evolution in furniture, where all its properties have been imaginatively exploited. This is particularly the case with the chair, which has always been a challenge for designers. Its function is more complex than that of any other item of furniture, because it must accommodate the human form in all its variety of sizes and shapes, sitting postures and positions, as well as suiting the physical and social environment. The historical development of the chair parallels technological advances and reflects cultural changes.

The history of GRP chairs really begins in 1940 with the competition sponsored by the Museum of Modern Art, New York, called 'Organic Design in Home Furnishings'. The 'organic' forms that became possible through the use of formed plywood and urea-formaldehyde glues encouraged new directions in design; up to that time chairs had been based on box-like structures and any bending of timber had been in one plane only, and not like the newly produced three-dimensional compound curved shells. Charles Eames and Eero Saarinen won first prize in the two main sections of the competition with moulded plywood chair shells, providing a useful foundation for the development of GRP designs.

The DAR chair, designed in 1948, was the first self-supporting one-piece GRP chair shell, the form of so many later plastics chairs. It was designed by a team including Eames' wife, Ray, for the 1948 'International Competition for Low-cost Furniture Design' also organized by the Museum of Modern Art, New York, where it shared

above
Tulip armchair designed by Eero Saarinen
in 1956, manufactured by Form
International. GRP shell on cast aluminium
base coated with nylon. 66 × 59 × 81cm
(26 × 23¼ × 31⅞in.(. Saarinen's famous
statement, 'I want to clear up the slum of
legs', was a reaction to the post-war
proliferation of steel legs and struts. (Photo
John Rose and John Dyble.)

DAR armchair on base of steel struts,
designed by Charles Eames in 1948. The
first self-supporting GRP chair shell.
64 × 61 × 82cm (25¼ × 24 × 32¼in.).

second prize with an early inflatable by David Pratt. The shell, on a frame of steel
struts, was made in the austere and restrained colours reminiscent of wartime
vehicles and aircraft—'gun metal, medium grey, off-white and light grey-brown'—
with the fibre reinforcement visible. The following year a side chair, the 'DSS' chair,
was produced, hot-pressed to obtain a good surface finish on both sides.

Whereas Eames' curved plastic chair shell was fixed onto a contrasting metal base,
Saarinen's 'Tulip' chair was the first design to produce a complete, flowing shape in
plastic, although the base was actually cast aluminium coated with white nylon. The
classic beauty of this form set the pattern for many chairs and tables now to be found
in every high-street furniture store. The two parts of the chair flow together and
appear to grow like the flower of its name in a homogeneous unity. Where Eames'
chairs are assertive and uncompromising, Saarinen's designs are always softened,
whether by gentler lines or by the use of upholstery. Saarinen said in 1957 that the
chair 'should not only look well as a piece of sculpture in the room, when no-one is in
it, it should also be a flattering background when someone is in it—especially a
female occupant'.

DSS side chair, designed by Charles Eames in 1949 to complement the DAR chair. GRP shell hot-pressed in matched-metal moulds fixed to legs of steel rod with rubber feet on self-levelling glides. Both chairs are manufactured by Herman Miller and moulded in a variety of bright colours, with or without upholstery and on different types of base.

CL9 (Ribbon Chair) designed by Cesare Leonardi and Franca Stagi in 1969, manufactured by Elco, Bellato. 99 × 67 × 60cm (39 × 26⅜ × 23⅝in.). A different solution to the problem of attaching a GRP form to a metal base. Here, a continuous band of GRP on a steel tube base exploits the tensile properties of the material.

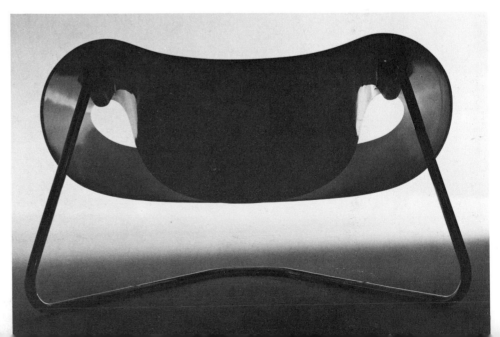

Prototype of the first plastics bicycle in 'Lexan' polycarbonate structural foam designed by Charles Cadorette and Joseph Dorrity in 1973. Manufactured by The Original Plastics Bike Co. Inc. New York. All the metal components of the bicycle have been replaced by polycarbonate structural foam except for the cable and shoes which are standard materials. *(Materie Plastiche ed Elastomeri)*

below
Tableau by Edward Keinholz. GRP figure in real car with Coca-Cola tin, bottle and paper cup. Keinholz began his 'tableaux' in the late 1950s. They were full-size petrified happenings, often luridly dramatic. Where Gilardi's hyper-realism seems romantic, Keinholz's imagery illustrates the tragedies of modern American Life. *(Materie Plastiche ed Elastomeri)*.

Homme (Man) designed by Ruth Francken in 1970 in Paris. GRP on chromed metal base. (Photo by courtesy of the artist.)

The headless 'Homme' chair in glossy GRP shows another way of fitting a seat over a metal frame, this time exploiting the cantilever in man's natural sitting position with the thighs smoothed out to form a flat lap.

The first complete single moulded chair in commercial production appeared in 1960, designed by the Danish architect and designer Verner Panton. An attempt was made to mould the shape in rigid polyurethane foam, but owing to the surface-finishing problems of the material it was manufactured instead in GRP by Fehlbaum, West Germany. As demand for the chair has increased, it has been produced in

138

Timber chair with lady and cats made by Rupert Carabin in 1896, with the appearance of a modern plastics moulding. The smoothness of the nude and cats contrasts with the chisel-marked seat. (Photo Laurent Silly Jaulmes)

above

Examples of urea-formaldehyde mouldings, including: cream-coloured 'Bandalasta Ware' sandwich box, beakers and plates from a 'Coracle' picnic set, 1927-32; exploding 'Rolinx' cigarette box; armchair tray with spring-loaded flaps; round box with lady and dog moulded in relief, 1947, designed by Harry J. Earland for Pignon Plastics Ltd; green 'Beetleware' picnic cups and plates; black and white 'Acme' vacuum flask; green 'Thermos' flask in wood-filled urea; also 'Quickmix' egg-flip shakers, biscuit bucket, pink cotton wool dispenser, yellow sugar shaker, 'Velos' napkin rings, box with moulded relief of the huntress Diana, and early salt and pepper pots of 'Beetle' resin. (Author's collection and objects lent by Dr and Mrs Nicholas Kemp. Photo Fritz Curzon)

opposite

'Bandalasta Ware' by Brookes and Adams Ltd, Birmingham, moulded in 'Beetle' thiourea-formaldehyde resin from Beetle Products Ltd, 1927-32. The photograph shows: four picnic horns; an imitation tortoiseshell box in blow-moulded celluloid with plated hinge, catch and label; and a fruit or rose bowl (the base of this screws into a threaded brass insert in the bowl). (Author's collection. Photo Ken Randall).

materials suitable for larger runs. Around 1968 it was made in 'Luran S' (styrene-acrylonitrile by BASF) by Fehlbaum; in 1974 it appeared injection-moulded in glass-filled nylon by Herman Miller; and it is now available in ASA terpolymer (acrylonitrile styrene-acrylic rubber, a tough new thermoplastic).

The Panton chair adopts the cantilever principle, originally developed in tube metal by Mart Stam and Marcel Breuer at the Bauhaus, now translated entirely into plastics and expressing the properties and qualities of GRP in its rounded and flowing form. All the GRP cantilevered chairs that followed possess the same graceful silhouette, expressing the dynamics that characterize plastics. Although called a 'stacking' chair it is more correctly a 'nesting' chair.

GRP lends itself to even more imaginative and bizarre forms, exemplified by the 'Floris' furniture, a range produced by Günter Beltzig and his brothers, of Wuppertal, West Germany. The chair illustrated is an ergonomic and anthropometric exercise in support, in the manner of a dentist's chair—a skeletal 'space age' structure designed to hold the seat, hips, spine and neck of suitably sized people. The seat and back have a channel down the centre—a structural detail which allows air to circulate.

above
Chair from the 'Floris' range, Beltzig Brothers, Wuppertal, Germany. Moulded GRP. (Maples Ltd)

Cantilevered chair designed by Verner Panton in 1960, manufactured by Herman Miller International. Originally moulded in GRP, now in ASA terpolymer. 49 × 57 × 82cm (19¼ × 22½ × 32¼in.).

Selene chairs and table designed by Vico Magistretti in 1968, manufactured by Artemide. Originally moulded in GRP, now ABS. The fluted leg sections extenuate and lighten the design. (Photo Aldo Ballo).

In contrast to the sculptural one-piece chairs, cantilevered like Panton's, cube-shaped like Rodolfo Bonetto's 'Melaina' chair (Driade, 1970) or hollow like Eero Aarnio's designs, the traditional four-legged chair can be successfully made in fibre-reinforced materials. One of the first in production was Pierre Paulin's 'Chair 300' (1965-66). More traditional in form, moulded first in GRP and later in ABS, is the classic simplicity of Vico Magistretti's 'Selene' stacking chair.

The translucent and heat-resistant properties of GRP can be successfully used to make lamps and moulded light fittings, a relatively unexploited field. Ettore Sottsass Jr designed a wide range of undulating mouldings in 1971 to make illuminated Odeon-style mirror frames, bed heads, lamps and cabinets. In 1974 GRP lampshades appeared looking like flapping handkerchiefs.

For many years large-scale mouldings in GRP were underdeveloped because they relied on craft-based processes. Only a few pieces appeared in the fifties and sixties which explored the potential of the material for interior design, although the production of boat hulls, customized furniture and metallic-flake car bodies continued as usual. Since 1972, however, improved processes have allowed GRP, with its superior properties for outdoor use, to become economically viable for larger-scale application in architecture. In furniture functional elements began to be combined into larger units. Chairs and tables were moulded together to fit out cafés and restaurants. The complete moulded bathroom was a logical development of this aggrandizement.

GRP architectural mouldings are now mushrooming; doric and fluted pilasters, door reveals and window pediments, and all the necessary components for 'luxury' dwellings' and 'commercial properties'. Often they have a valid function,

above Measuring the tension stresses in a loaded model of the Casarino children's chair. *below* The finished chairs in ABS under real stress. (BASF)

particularly in the renovation of old houses; it is far safer to replace a cornice in GRP, which is easy to fix, maintenance-free, and weather-resistant, then to re-make (labour-intensively) the rendered mouldings, which in a few years would need repair again, or if left alone would flake, crack and possibly fall. However, they also appear as an architectural cliché in 'architect-designed houses' in mock-Regency style, with doors and porticos slapped on with no visual relationship to the architectural properties of the building.

Fortunately this trend in pastiche is counter-balanced by the rapidly increasing use of GRP cladding panels for commercial buildings, where the material is used in its own right. The pipes, sand, bricks, cement and shuttering used in underground drainage works are also being replaced by GRP pultrusions and interceptor tanks moulded in one operation.

Bouloum chaise longue, designed by Olivier Mourgue in 1968 for the main hall of the French Pavilion, Osaka Expo. GRP shell upholstered with flexible foam and stretch jersey. (Airborne)

145

The first all-plastics house designed by Ionel Schein, with R. A. Coulon and Yves Magnant in 1955-6 for the Salon des Arts Ménagers, Paris 1956. Contains a GRP bathroom The house now stands in the grounds of the Charbonnages de France, Douai.

In the application of glass fibre to bathrooms, many self-contained or component units have been designed but these rarely fit in with new or existing structures, which usually have immensely variable dimensions. The only way of coping with this problem has been to provide bathrooms with the elements separate, to be arranged as desired, and leaving the selection of finishes to personal taste.

146

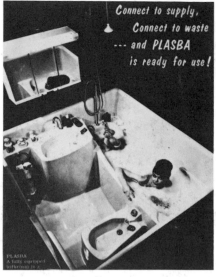

above

PLASBA moulded bathroom by the Société PLASBA, Paris. Manufactured in the mid 1960s as a single moulded unit as shown, or minus two walls, and in two reduced versions. As a plug-in unit the size is limited to 165.1 × 177.8cm (65 × 70in.), but with a height of only 104.1cm (41in.) (note the mirrored cabinet and shower suspended in mid-air in the photograph!). It can blend into any room, disguised with plants, shelves or tiles. But the difficulty of moulding undercuts resulted in no foot space beneath the bath and basin. Weight 165lbs. (74.9Kg).

GRP bathroom units designed by Alberto Rosselli in collaboration with the Montecatini Plastics Research Institute at Castellanza in 1957. Rosselli's first experimental design in plastics, intended for mass-production. Separate panel-moulded GRP units carry a pre-fabricated plumbing system on their reverse side, ready for connection. The units fit flush together providing greater flexibility than Schein's curved bathroom. (Photo Publifoto)

Polyethylene Terephthalate (PETP) *thermoplastic*

Polyethylene terephthalate (PETP) is an 'improved' polyester, used mainly as film and fibre under some well-known trade-names such as 'Terylene' and 'Dacron' fibres, and 'Melinex', 'Merolite' and 'Mylar' transparent film. It is a linear thermoplastic polyester produced by the condensation reaction of ethylene or propylene glycol and an unsaturated dibasic acid.

PETP fibre is almost as strong as nylon, and can be hot-drawn through a spinneret. It was discovered in 1941 by Whinfield and Dickson at the Calico Printers Association, who continued the early work done by Carothers in America, and was developed by ICI under the trade-name 'Terylene', and produced commercially in America by Du Pont in 1953 under the name 'Dacron'. It has enjoyed great success

Clouds by Andy Warhol, 1966. Aluminium-coated 'Mylar' polyester film filled with helium. Very early inflatables, they were made for the exhibition at the Leo Castelli Gallery, New York, to mark the end of Warhol's period of painting. The idea was to let them out of the window as floating paintings. (Leo Castelli Gallery, New York)

on account of its crease and fungus resistance, which makes it ideal for 'drip-dry', non-iron clothes and furnishings.

In the fifties twisting and crimping yarns were developed and as PETP is thermoplastic it could be heat-softened and moulded at a temperature of around 230°C. This process instigated the fifties' fashion for pleated skirts, made of blended wool and polyester. More recently crimped fibre has appeared as a substitute for feathers and down in quilts and duvets. It had already been used in upholstery as a soft top layer over denser foam, the fibres making an ultra-soft plump white cloud, shaped or stitched into lengths of piping ready for machining.

Polyester fibre is used for industrial sheeting and drive belts, and cord for making car tyres and marine ropes. 'Terylene' safety nets were used while building the Forth Road Bridge, and 'Dacron' textiles have been developed for use in surgical implants, much of the work being carried out at the Dow Corning Center for Aid to Medical Research, Michigan, along with investigations into 'Teflon' and the silicones.

Although polyester textiles feel very soft, they are in fact woven from hard fibres. Solid thermoplastic polyester can be moulded by injection and extrusion processes into very strong, nylon-like gears and bearings, door handles and tableware.

Polyester film is a strong, glassy but flexible film with good chemical and electrical

148

properties. It is oriented during manufacture, making it particularly strong and stable. The film is used in drawing offices, in photography as a film base, in the manufacture of magnetic tapes, and widely in packaging. One example, with demonstrates the toughness of the material and its impermeability to carbonated drinks is the American Pepsi-Cola disposable bottle, blow-moulded in Du Pont's 'Dalar' PETP film.

Polyester film can be coated with a fine aluminium deposit, about five millionths of an inch thick, by the vacuum metallizing process. Coated film is used bonded to vinyl sheet to make self-adhesive, tough, decorative trim for cars, radios, furniture, packaging, shoes and wall coverings. It can be high-frequency-welded to other plastics. Held up to the light, metallized polyester film is just transparent and has been used since 1972 for making solar screens and blinds. Laminated to PVC and perforated to permit an air-flow for ventilation, the metallic side reflects the sun's rays, but allows one a view out through it like a two-way mirror. A window blind made in this way can block 80 per cent of the sun's rays. The Echo satellites were covered with 'Mylar' polyester film.

UK: Beetle polyester, Beetle DMC and SMC (BIP), Catabond (Catalin Ltd), Cellobond (BP Chemicals International), Crimplene (ICI), Crystic (Scott Bader Co.), Deroton (ICI), Fibreglass (Fibreglass), Filon (BIP), Hydroglas and Hypul (BTR Reinforced Plastics), Melinex and Merolite (ICI), Metalon and Sunscreen (Chamberlain Plastics), Polyite (Minerva Dental Labs.), Terylene (ICI)
USA: Celaner and Celanex (Celanex Corporation of America), Dacron and Dalar (Du Pont), Fibreglas (Owens-Corning Fibreglas Corporation), Genpol (General Tire and Rubber Co.), Glaskyd (American Cyanamid Co.), Hetron (Durez Division, Hooker Chemical Corporation), Laminac (American Cyanamid Co.), Mylar (Du Pont), Polyite (Reichhold Chemicals), Premi-Glas (Premix), Trevira (Hoechst Fibres), Valox (General Electric Co.), Vibrin (Naugatuck Chemical Division of U.S. Rubber Co.)
ITALY: Gabraster and Respol (Montedison), Sniatron and Sniavitria (SNIA)
FRANCE: Rhodester and Stratyl (Rhône-Progil Société)
JAPAN: Ultrasuede (60% polyester fibre, 40% non-fibrous polyurethane, Toray Industries)
GERMANY: Alkydal (Bayer), Hostadur (Hoechst), Legumat and Leguval (Bayer), Roskydal (Bayer), Vestan (Bayer)

EPOXIDE RESINS (EP) *thermosetting*

The reaction of two petroleum derivatives, bisphenol A and epichlorohydrin, produces a thermoplastic chain capable of being cross-linked to form exceedingly tough epoxide, or epoxy, resins. These are highly resistant to heat and chemicals, and are mainly used as protective surface coatings, for which they were developed in the forties.

Epoxy is applied as a coating to the insides of tins of food and drink; it is also electrostatically sprayed onto metal window-frames, incorporated as a skid-resistant surface on motorways, and commonly emulsified as an interior and exterior water-based paint. Epoxy resin makes excellent waterproof coatings even for concrete and masonry.

Although more expensive than polyester, it is also used to strengthen fibres such as glass and asbestos. It is moulded by the same processes as glass-fibre-reinforced plastics.

Epoxy resins are especially useful for air and space travel, where very lightweight materials with first-class electrical properties are required. The eleven-foot-long beak of Concorde is made from reinforced epoxy resin. The heat shields on the Apollo space crafts incorporated epoxy resins, while electrical components in rockets and satellites are coated with it for protection at high temperatures. It has proved compatible with the human body, and has been used in Britain to encapsulate a nuclear pacemaker, which is powered by less than one fifth of a gramme of plutonium oxide.

Epoxy resin possesses remarkable strength as an adhesive for glueing a wide variety of materials: metals to glass and plastics, and even concrete to concrete, where it can make a bond look more like a welded seam than a glue-line. It is used in dental fillings. In most cases it is stronger than the materials being bonded.

The versatility of epoxy resin is shown by Frank Gallo's *Nude*. It was first sculpted in clay and then a mould was made from which an epoxide casting was taken. It was painted with coloured resin, then polished and textured. To produce the ivory-like glow of the skin, and to intensify the colours, Gallo used an acetylene torch to singe the surface of his model.

Foam Reservoir Moulding (FRM) (Elastic Reservoir Moulding)
A recent process patented by Shell, FRM produces strong, rigid but lightweight composite mouldings. A sandwich of flexible polyurethane foam between layers of glass reinforcement is placed in a heated mould. The foam has previously been impregnated with Shell 'Epikote' epoxy resin and 'Epikure' curing agent. When compressed inside the shaped mould the resin is forced out of the foam and into the glass matting, where it cures, forming, for example, a chair shell.

Glass-Fibre-Reinforced Cement (GRC)
Street furniture, planters, cladding and ducting are appearing in mouldings of glass-fibre-reinforced cement (GRC). A matrix of cement is reinforced with special alkali-resistant glass fibres, and impregnated with epoxy or polyester resin. 'Thycrete' GRC made by Thyssen has a hard vandal-resistant surface finish, and can be either brightly coloured or a very convincing 'marble'. It feels cold like marble, and even has that hard, stone-like ring to it.

UK: Araldite (Ciba-Geigy), Epikote and Epikure (Shell Chemicals), Polybond paints (Polybond), Hydrepoxy (Unibond), Thycrete (Thyssen)
USA: Epi-Rez (Davoe and Raynolds Co.), Epolite (Rezolin Inc.), Epon (Shell Chemical Co.), Unox (Union Carbide Corporation)

Nude by Frank Gallo. Epoxide casting from clay model, polished, coloured and textured. (Graham Gallery, New York)

POLYPROPYLENE (PP) *thermoplastic*

Once ethylene had been successfully polymerized into high-density polyethylene by Professor Ziegler in 1953, efforts were directed towards making a useful plastic material out of propylene, which had so far evaded the chemists. Like ethylene, propylene is a feedstock derived from petroleum. Also like ethylene, it is an olefin made from carbon and hydrogen ($CH_2 = CHCH_3$).

The breakthrough came in 1954 when Professor Giulio Natta of the Polytechnic Institute of Industrial Chemistry, Milan, used a Ziegler-type catalyst and polymerized propylene. He discovered that Ziegler's catalyst could be used to create a polymer with a specific molecular structure, the first time that a material had been made to order. Previously propylene had not been successfully polymerized because its molecular structure has a disorganized sequence (called atactic, from the Greek word meaning 'without order'). Natta found that the Ziegler catalyst orientated the molecules in a particular direction before polymerization, and produced two types of ordered arrangements, known as isotactic (from the Greek: 'in the same order'), and syndiotactic ('in contrasting order').

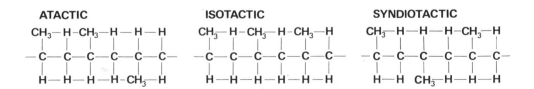

The three different chain structures of polypropylene determined by Natta.

Natta's pioneering work laid the foundations for the development of polymers with specified properties, and his discovery made possible the development of many new plastics. Predetermined reactions using the Ziegler catalyst system are called stereospecific.

Natta's original polypropylene was produced by Montecatini (now Montedison) under the trade-name 'Moplen'. Polypropylene can be produced by the same plant that produces high-density polythene, with only slight modification. As it is thermoplastic it can be moulded by standard thermoplastic processes: injection moulding, thermoforming, extrusion, blow and rotational moulding. It can also be heat-welded, and used as a coating. Because it degrades easily it must be compounded with stabilizers and anti-oxidants.

Polypropylene is a highly versatile material. Its impact resistance and flexional strength make it suitable for chair shells (for example in the ubiquitous Hille stacking chair), suitcases, TV cabinets, toys and bricklayers' hods. Although it has a waxy feel like polythene, it cannot be scratched as easily. It is the lightest of all the thermoplastics, and possibly destined to overtake polythene in future tonnage.

Polypropylene is extruded into pipes, film and sheet. The artist Christo wrapped a rocky Australian coast in one million feet of polypropylene sheeting. Biaxially orientated polypropylene film has been stretched both ways and is very tough. Laminated onto book jackets it can be bent thousands of times without cracking; and used as windows in envelopes and boxes it forms a moisture- and grease-resistant protection. Polypropylene has a higher melting point than polythene; it can be sterilized and thus has many useful home and medical applications: tweezers and dishes, filters for artificial kidneys, boilable Christmas pudding basins and baby-care products. Polypropylene foil is made into 'boil-in-the-bag' packets. Components can be chromium-plated for cars and for the insides of dishwashers and washing machines.

A unique property of polypropylene is that it can be moulded into integral hinges which can be flexed millions of times on briefcases, suitcases and tool boxes.

Polypropylene fibres and filaments can be melt-spun. They are stretched (orientated) and annealed, and made into straps for packing pallets, woven into shiny sacks, ropes and twine, made into pile carpets and carpet backing, and woven into fabrics. Polypropylene seaweed is even helping to prevent erosion along the New Jersey and Dutch coasts. Like many other plastics moulded polypropylene can be reinforced with glass fibre to increase its strength and heat resistance; with this treatment it forms the distributor belt guard in the Fiat 128, and the 'Propathene' air grille on the Lotus Elite. It can also be reinforced with asbestos, again used for car components, and with talc, for car instrument boards or the detergent compartments in dish-washers and washing machines.

An even tougher grade of polypropylene can be produced by copolymerization with a small quantity of ethylene monomer. This copolymer functions better at lower temperatures where other plastics might crack, and is used for yoghurt pots, ice-cream containers and beer crates.

Almost ten years passed from Natta's polymerization of propylene in 1954 before the material was used in furniture. The polypropylene stacking chair appeared on the market in 1963 designed by Robin Day for Hille and Company (now Hille International), and it became the most famous plastic chair in the world. Day's chair is now made under licence in fifty countries, which keeps the price competitive (over six million have been manufactured), and has spawned many imitators, some of whom have been taken to court by Hille.

It represented a breakthrough in furniture design. It was the first chair to be injection-moulded, the precursor of all mass-produced plastics chairs; it was the first chair specifically designed for production in the new thermoplastic polymer, polypropylene; it was also the first single injection-moulding of a self-supporting chair shell. Eames' stacking chair had been laminated from GRP, but there are many

Polypropylene side and armchair designed by Robin Day in 1963, manufactured by Hille and Co. Ltd. Injection-moulded in 'Shell' polypropylene on straight tubular metal bases. The first injection-moulded chair shell, 'often copied, never bettered'. (Photo Dennis Hooker)

opposite
Two-part injection mould for the Hille Polypropylene Chair, made by Talbot Ponsonby Ltd, moulding carried out by Thermo Plastics Ltd.

similarities of form—the rolled-over edges for extra strength, and the smooth flowing body-contoured shape.

Polypropylene was selected because it was much cheaper than GRP and highly suitable for mass-production. The aim was to design an all-purpose chair at the lowest possible price, and this intention was certainly achieved. When it was launched, the cheapest stacking version cost only £2 19s. 6d. (or just over 8 dollars at then contemporary rates of exchange). It has a grained imitation leather surface which prevents the sitter from sliding about, such as is found on polypropylene suitcases and children's steps. However, the chair contravened the then current British Standard recommendation that a gap should be left between the seat and the lumbar support.

Polypropylene Structural Foam (PP SF)
Polypropylene structural foam was pioneered by Shell Chemicals and possesses all the properties of polypropylene already mentioned but with greater rigidity. In solid polypropylene extra strength is obtained from reinforcing-ribs or by increasing the

Battery cover for Bedford heavy commercial vehicle, injection-moulded in ICI 'Propathene' polypropylene structural foam by Cabinet Industries Ltd in 1975. 50.5 × 41.5 × 25.2cm (20 × 16¼ × 10in.) yet weighs only 2.6 Kg. (ICI)

thickness. In structural foam it is the nature of the material itself—a dense surface with an inner core of foamed plastics—that provides the increased strength. It has the swirling surface pattern characteristic of structural foam, although it is often painted over. Various grades of polypropylene structural foam are produced for injection moulding and casting. Among the applications of this material are the moulding of canoe paddles, and an extremely diverse collection of specially designed containers and trays, for transporting everything from vegetables straight from the fields, to bottles and tools.

UK: Bexphane (BXL), Carlona P (Shell Chemicals), Propafilm and Propathene (ICI), Shell Polypropylene (Shell Chemicals), Ulstron (ICI), Vulcathene (Stadium)
USA: Marlex (Phillips Petroleum Co.), Oleform (Avisun Corporation), Pro-fax (Hercules), Prolene (Industrial Rayon Corporation)
GERMANY: Novolen (BASF)
ITALY: Kastilene (ANIC), Moplen (Montedison)

ACRYLONITRILE-BUTADIENE-STYRENE (ABS) *thermoplastic*

ABS resin is a terpolymer made up from three petroleum-derived monomers, acrylonitrile, butadiene and styrene. When copolymerized with acryonitrile alone, styrene produces a strong material with good chemical resistance, styrene-acrylonitrile (SAN). In the late forties it was discovered that if styrene-acrylonitrile was polymerized with butadiene in the form of fine particles of modified rubber, not only was the chemical resistance further improved, but the added butadiene increased its impact resistance. By varying the proportions of the three ingredients, as well as altering the molecular arrangement within the polymer, ABS resins can be produced with a variety of properties to suit different processes and functions: high impact, extra-high impact, medium, and low impact, heat-resistant, self-extinguishing. A typical ratio of the three constituents is acrylonitrile, 20 per cent; butadiene, 20 per cent; and styrene 60 per cent. ABS degrades in ultra-violet light, but this can be countered by laminating acrylic film onto sheet for thermoforming, or by painting or plating mouldings.

Early ABS resins were calendered into sheets or extruded as piping, although they could also be injection-moulded and blow-moulded. The sheets were very expensive to produce and were not commercially viable until 1948. ABS sheet is best suited to thermoforming and is vacuum-formed into a wide variety of articles. In 1965 one of the earliest ABS chairs was vacuum-formed from 'Vulkide A' by James Nuttal Limited. Supported by tubular metal frames, the chairs could be stacked twenty high. Refrigerator door liners and wash basins are vacuum-formed from ABS, and all sorts of shelving units, such as the 'Brick system' devised by De Pas, D'Urbino and Lomazzi for Longato (1972). Dorothee Maurer-Becker's 'Wall-All' wall hanger, for Design M, Munich, is a one-piece moulding with shaped rigid pockets in which to store an assortment of odds and ends. Ettore Sottass Jr used vacuum-formed ABS for the luminous parts of neon-lit wall mirrors. Interior car-door panels have been thermoformed in ABS, and entire car bodies such as the Citroën Méhari.

ABS was the first plastic to imitate metals convincingly and can be chromium-plated, not an easy process with plastics. Such mouldings provide lightweight replacements for electroplated metal parts on cars and cameras, for example.

Nike chaise longue designed by Richard Neagle and E. Szego in 1968, manufactured by Sormani, Como. Vacuum-formed ABS sheet on swivel base of enamelled steel. Lower part of the seat is filled with rigid polyurethane foam. Illustrates strengthening ribs on back and the shape limitations of the vacuum forming process. (Design Council)

ABS is pleasantly warm to the touch and many domestic items with a high-gloss finish have been moulded in it: housings for electrical goods such as hairdryers, shavers, torches and vacuum-cleaners, food mixers and toys such as 'Lego'. The injection-moulded acrylic Post Office telephones of the fifties proved to be too brittle and in 1963 were changed to ABS. The hand-sets of these later telephones are half the weight of the early phenolic models, and can in some cases be too light.

A classic domestic appliance moulded from ABS is the Terraillon BA 2000 kitchen scale, which was almost immediately selected for the permanent collection of the Museum of Modern Art, New York. It was designed by Marco Zanuso, who had been commissioned by the Terraillon Corporation in France to design first the TI 11 bathroom scale, and then the BA 2000 kitchen scale, a fitting weighing-machine for the land of gourmets. It is one of France's best-sellers; by 1974 378,000 had been sold. It is very neat and compact in design, but also possesses an aura of mystery. The flat-topped lid, which is machine-washable, can be used for large dry objects, even letters, or, turned upside down and fitted to the base, it can be used to measure liquids and powdered solids. The weight indicator is magnified and angled to make it easy to read from a working position.

The first one-piece injection-moulded chair was moulded in ABS, to a design by Joe Colombo, his 'Seggio' 4860. Although it was designed in 1965, it was not produced commercially until 1967, the same year that Verner Panton's one-piece cantilever chair appeared in hot-pressed GRP. Unlike Panton's cantilevered chair, this design is far more traditional with a seat, back and four legs. Seen against many later and more slender plastic chairs, it appears heavy and squat, but its generous dimensions assure comfort and unfussy solidity.

Seggio 4860 designed by Joe Colombo in 1965 by Kartell, Italy. The first entire injection-moulded chair, initially in 'Cycolac' ABS, later in 1971 in Bayer's 'Durethan' nylon which was the first chair with a 5-year guarantee. Choice of two heights of attachable legs. Described as a 'stacking' chair, yet only manages it in threes. (Abitare)

ancora una volta la prima

la prima ad essere stampata interamente
in materiale plastico.
questa sedia è ora la prima
ad essere prodotta in resina
poliammidica ed è ancora la prima
ad essere venduta con una garanzia
di 5 anni, non invecchia, non si rompe.
può essere lanciata dalla finestra,
lasciata all'aperto, immersa nell'acqua,
portata al polo o nel deserto
e rimane sempre nuova.

Kartell è design

disegnata da Joe Colombo e prodotta in Durethan Bayer dalla Kartell spa di Binasco (Milano)

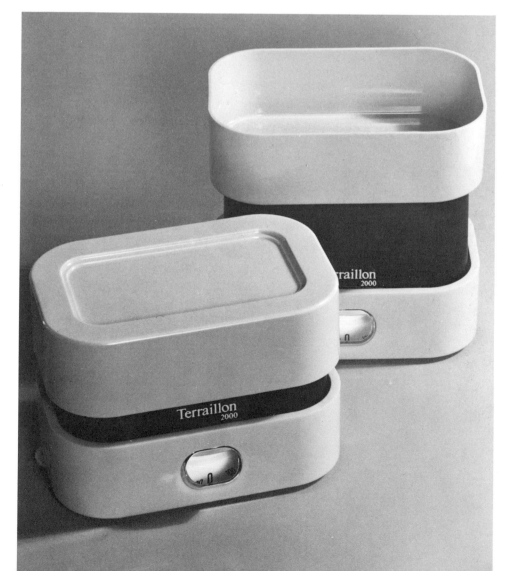

above
The Input Collection designed by Martin
Roberts of Conran Associates, manufactured
by Crayonne Ltd (Airfix Plastics Group) in
1974. One of the most comprehensive ranges
of ABS containers, consisting of twenty-one
items. The ash-tray has a toughened
melamine liner and the rubbish bin has a
rubber lid. The most controversial feature is
the use of heavy-duty ABS in a domestic
situation at a time when plastics mouldings
are trying to save material. The range brings
back the substantial and pleasurable feel of
early phenolic mouldings.

Terraillon BA 2000 kitchen scale designed
by Marco Zanuso, manufactured by
Terraillon Corporation, France. Many
people are not sure what this is when packed
away. The flat-topped lid is turned upside
down for measuring liquids or powdered
solids.

Marema nest of four ABS stacking tables designed by Gianfranco Frattini in 1968, manufactured by Cassina. The legs have been given vertical strengthening ribs, the only decorative detail on an otherwise simple design. Stacked on top of one another, they can form a tallboy. (Photo Aldo Ballo)

By the end of the sixties furniture components were becoming very sophisticated and 'engineered'. Components for entire ranges of shelving units were moulded in ABS. Not only cabinets but tables and chairs benefited from the component system which has accompanied recent plastics design. It facilitates transport and packaging, and also the replacement of damaged parts. The restrained and unstylized 'Desco' design illustrates how plastics technology has been adapted very easily to traditional forms.

In the construction industry ABS competes with high-density polythene, unplasticized PVC and polypropylene in the replacement of cast-iron piping and fittings. Moulded out of ABS, the 'Osma' underground manhole, hydrodynamically designed to aid the flow of water, looks like a work of art.

UK: Abstrene (BXL), Cycolac (Borg-Warner Chemicals), Kralastic and Royalite (Uniroyal), Sternite (Sterling Moulding Materials), Vulkide A (ICI (Hyde))
USA: Cycolac (Marbon, Borg-Warner Corporation), Kralastic (US Rubber Co.), Lustran (Monsanto)
GERMANY: Novodur (Bayer)
ITALY: Cycolac (Borg-Warner Chemicals), Ravikral (ANIC) Urtal (Montedison)

Multiplo-Proposta 60 shelving designed by Rodolfo Bonetto, manufactured by Harvey Guzzini. Injection-moulded units in 'Ravikral' ABS, the first ABS furniture system to incorporate cupboard doors. The supports are double-skinned and have connectors located inside for building up height. The permutations possible are as varied as the composer's imagination and necessity. (Victor Mann and Co. Ltd. Photo Simion)

Desco seats and table designed by Carlo Hauner, manufactured by Elco, Italy. Injection-moulded in 'Ravikral' ABS. Both chair and table knock down into three components. Particularly interesting is the corner support for the table with its lattice of ribs and webs like the vaults of a miniature Gothic Cathedral. Legs are integrally moulded to take screw fittings.

POLYCARBONATE (PC) *thermoplastic*

In the early days of plastics, choosing a suitable material for a particular job from the few that were available, normally either a thermoplastic or a thermoset, was a relatively easy task. Each polymer had its defined properties and, as far as the layman was concerned, each plastic material was easy to recognize. It is now becoming much more difficult as plastics have become increasingly complex. Since polypropylene, the first truly made-to-measure polymer, there has been more and more cross-breeding, producing plastics not only with superior properties, but with properties that overlap, thus making the chemist's and designer's choice more exacting. Often the material chosen is the result of compromise between the most suitable plastics and the financial resources available.

Polycarbonate resins are one of a new breed of engineering plastics, mostly thermoplastic, which are used where tough materials are required, usually as a replacement of metal. These high-molecular-weight materials include the mould-able polyesters, polysulphone, polyacetal and polyphenylene oxide. Polycarbonate resins are long-chain thermoplastics made from bisphenol A and either carbon dioxide, carbonyl chloride, or diphenyl carbonate, and are transparent.

The repeating polycarbonate chain unit

Although cross-linked polycarbonate was discovered in Germany in 1898 by Einhorn, another half-century passed before Farbenfabriken Bayer produced the first thermoplastic moulding polycarbonate in 1953. Bayer put 'Makrolon' into commercial production in 1959, and this was followed by the American version 'Lexan', produced by the General Electric Company, who had independently developed polycarbonate at the same time.

Polycarbonate is extremely strong, and in sheet form is 250 times stronger than laminated glass. Its rigidity and weather and impact resistance are retained even at low temperatures, making it ideal for sports equipment such as safety helmets, goggles and badminton rackets. Being transparent as well as tough makes it one of the most vandal-proof materials available for football pavilions, street lighting, telephone kiosks, car lamps, gymnasium glazing. In Northern Ireland 'Makrolon' sheet is incorporated into personnel vehicles and forms the visors and shields which protect British troops.

Despite all this polycarbonate tends to suffer from stress-cracking if it comes into contact with detergent, but being non-toxic as well as strong it is moulded into transparent, impact-resistant babies' feeding bottles and tinted tableware. Owing to its high softening point, and dimensional stability, it is used for precision components for cameras and projectors, hairdryers and shavers, coffee grinders, percolators and hair curlers. On catching fire it bubbles, but can be made flame-retardant and self-extinguishing.

Polycarbonate can also be reinforced with glass fibre like many plastics to increase its impact strength and heat resistance. Unfilled polycarbonate softens at $135°C$ but glass-filled polycarbonate does not begin to distort until $140-145$ C.

Moulding Processes

Polycarbonate resin in granular form can be moulded by standard thermoplastic processing equipment. It can be blow-moulded to form light diffusers or extruded into pipes, rods, profiles, film and sheet. In powder form it can be cast, or used as a coating. Injection moulding of glass-filled polycarbonate is carried out in very hot moulds; the higher the mould temperature and injection speed, the glossier the

opposite
The Malibu Collection of furniture designed by Heinz Meier in 1972, manufactured by Landes Co., USA. A commercial use of ready-made ABS plumbing fittings with 'Tough Stuff' fabric of nyl

surface.

Polycarbonate sheet can be vacuum-formed and deep-drawn and easily machined. It can be cut and turned, bonded ultrasonically or with solvents, or riveted and screwed. It lends itself to snap-fit closures and the surface can be decorated with paint or metal coatings or printed, or simply timber-grained.

Polycarbonate Structural Foam (PC SF)

Like most thermoplastics, polycarbonate can be made as a structural foam, and is similar to polystyrene structural foam in its composition and appearance, but is much stronger, being the strongest of all the thermoplastic structural foam resins.

Both General Electric's 'Lexan' and Bayer's 'Makrolon' are available in a foaming grade. It is easy to cut, can be nailed like wood, and the visible parts can be painted, sprayed or metallized to cover the characteristic swirling pattern. The strength-to-weight ratio is double that of aluminium and five times that of steel, and together with far greater design freedom these advantages make structural foams an attractive alternative to die-cast and stamped metals, especially as it can be as much as 75 per cent cheaper. The 'Lexan' bicycle, launched in New York in February 1973, was the world's first production bicycle to be made almost entirely of plastics. It is corrosion-resistant, lubrication-free, and weighs only 7.24 kg (just under 16 lbs).

UK: Lexan (Engineering Polymers), Makrolon (M & B Plastics)
USA: Lexan (General Electric Co.), Merlon (Mobay Chemical Co.)
GERMANY: Makrolon (Farbenfabriken Bayer)

POLYACETALS (POM; POLYFORMALDEHYDES) *thermoplastic*

The polyacetal resins are a group of engineering thermoplastics designed to replace die-cast and stamped metals. They are usually referred to as acetal and are derived from the aldehydes. Formaldehyde (formalin) is used for making the phenol-, urea, and melamine-formaldehyde plastics. If formaldehyde itself is polymerized, a high molecular weight polymer results, polyformaldehyde, with the repeating chain link written as CH_2O or

$$-O-\underset{\underset{H}{|}}{\overset{\overset{H}{|}}{C}}-O-\underset{\underset{H}{|}}{\overset{\overset{H}{|}}{C}}-$$

First discovered in 1859, the pioneering work on polyformaldehydes was carried out by Hermann Staudinger in the 1920s and it was not until 1960 that E. I. Du Pont de Nemours succeeded in commercially producing a usable polyformaldehyde resin under the trade-name 'Delrin'. It was closely followed by 'Celcon', an acetal copolymer made by the Celanese Corporation.

Polyacetal is an intense white resin, very hard and tough, and heavier than most plastics. Almost all its functions are as a replacement for metal, not only the softer metals such as aluminium, brass and zinc, but also steel. In Germany acetal wood-screws have been moulded by Hoechst, and acetal can be found as door and refrigerator handles and as cranks on window frames. The ability of acetal to resist high temperatures (its melting point is 347°F) makes it a suitable replacement for machinery parts, such as bearings, high-speed gears, and fans for washing-machines and extractors. David Mellor's cutlery with ivory 'Delrin' handles is machine-washable. Acetal has all the qualities of a 'super-plastic' and is correspondingly expensive and not used in large quantities.

Acetal is usually injection-moulded, but it can also be blow-moulded (into aerosol containers for example) and extruded, as well as post-machined. Its low coefficient of friction together with its high tensile strength make it suitable for wheels and castors;

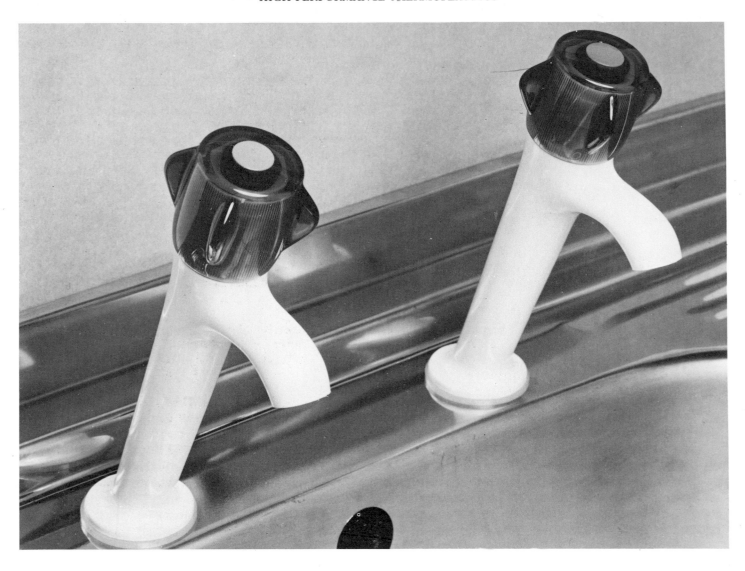

parts of the 'Triolite' push-chair moulded in acetal can withstand all the twisting that tends to pull other push-chairs apart.

UK: Kematal (ICI)
USA: Celcon and Kematal (Celanese Corporation), Delrin (Du Pont)
GERMANY: Hostaform C (Hoechst)

Opella 500 range of taps in 'Kematal' ICI acetal resin, designed by Martyn Rowlands manufactured by IMI Developments Ltd in 1969—the first plastics taps in the world, the first model was tested in 1966. Moulded acetal is so smooth that furring is eliminated, and the taps do not heat up, stain or get scratched.

HIGH-PERFORMANCE THERMOPLASTICS

Despite the general consumer trend towards a return to natural materials like timber, new types of plastics are still being developed and applied. These new 'super-plastics', which are mostly thermoplastic, have exceptional properties, such as very high softening points or high resistance to chemicals. Although they are not yet used in large quantities, they have specialized, high-performance uses, for example in components for machines, motor-cars, rockets and satellites. In these areas they carry out functions formerly performed only by metal.

Although many of these plastics are still in development and do not yet have an established history, some have already given proof of their performance, including chlorinated polyether, ionomer resins, methylpentene polymers (TPX), the thermo-plastic polyesters—polybutylene terephthalate (PBT), polyethylene terephthalate (PETP) and polytetramethylene terephthalate (PTMT)—polyphenylene oxide (PPO), polyphenylene sulphide (PPS), polysulphones (PES), and polyimides.

Chlorinated Polyether *thermoplastic*
Chlorinated polyether is a corrosion-proof and chemical-resistant material, used for tanks, piping, tubing, valves, wire-covering, tank-lining sheets or as a coating on metal.

USA: Penton (Hercules Incorporated).

Ionomers *thermoplastic*
The name ionomer covers a range of new tough flexible thermoplastic materials based on ethylene. They were developed in 1965 by Du Pont under the trade-name 'Surlyn'. As they can be made transparent, they are widely used as a tough vacuum-packaging film, and are sheet-formed into refrigerator trays, blow-moulded into bottles, and injection-moulded into transparent hammer-heads and tool-handles. 'Surlyn' is replacing gutta percha as a coating for golf balls.

UK: Alkathene Ionomer (ICI)
USA: Alathon (Du Pont), Fortiflex (Celanese Corporation), Marlex (Phillips Petroleum Co.), Surlyn (Du Pont)

Methylpentene Polymers (TPX) *thermoplastic*
Methylpentene polymers are transparent and used for laboratory and food containers, medical equipment such as tubes and syringes, and light fittings.

Polytetramethylene terephthalate (PTMT) *thermoplastic*
Polytetramethylene terephthalate already competes with polycarbonate, acetal and nylon. It is a thermoplastic polyester with a very low water absorption, hence its usefulness for pumps.

USA: Tenite PTMT (Eastman Chemical Products)
UK: Impol (ICI), Deroton (ICI)

Polyphenylene Oxide (PPO) *thermoplastic*
Polyphenylene oxide was developed by the General Electric Company in 1964, and under the trade-name 'Noryl' offers ten standard and three high-performance grades, including two flame-retardant grades. Like other engineering plastics it can be injection-moulded or extruded, either unfilled or glass-reinforced. In 1971 the technique for plating polyphenylene oxide was developed, and is used, for example, for shower-mixing knobs and automobile grilles. The heat-resistance of Noryl 110 is so high that it has been moulded into a transparent coffee percolator. PPO structural foam is a rigid foamed core within a very tough skin and is replacing metal for machine housings such as typewriters, cash registers, consoles and TV cabinets, where its high impact-resistance cuts down sound vibration. The Volkswagen Golf car has an instrument panel moulded in Noryl structural foam.

UK: Noryl (Engineering Polymers)
USA: Noryl (General Electric Company)

Polyphenylene Sulphide (PPS) *thermoplastic*
Polyphenylene sulphide is a very rigid tough material that softens at high temperatures. It crosses the traditional categories of plastics, for it is a thermoplastic that can be made to act like a thermoset. It can be injection-moulded or compression-moulded, and it can be reinforced with glass, to produce parts of car engines, for example, and non-stick kitchen implements.

USA: Ryton (Phillips Petroleum)

Business machine cover moulded in Noryl PPO structural foam showing lugs, bosses and reinforcing ribs moulded into the component. (Engineering Polymers Ltd)

Futura Forgettle automatic electric kettle designed by Julius Thalmann, manufactured by Russell Hobbs. Originally moulded in Noryl modified PPO in 1973, with a stainless steel top. It was the first electric kettle with the automatic switch unit incorporated into the body, only possible through the use of plastics, as was the integrated spout, through which it is filled. It does not heat up or attract scale. The 1978 model is moulded in Kematal acetal copolymer with a black phenolic handle.

Polysulphones *thermoplastic*

Polysulphones are workable up to 300°F (150°C) and are used for electrical components, machine housings, and automotive and aerospace equipment. 'Udel' by Union Carbide was the first to be developed, in 1965. It can be injection-moulded, extruded, blow-moulded and thermoformed, and is reported to be more economical than stainless steel.

Conclusion

To counteract prejudice against plastics, various sections of the industry have pleaded for many years for a wider understanding of synthetic materials. The consumer needs more guidance on the use and care of plastics, and should be advised of the standards of performance which can be expected.

Alan Glanvill's paper 'Plastics and Society', read at PINTEC 72 (Plastics Institute National Technical Conference), condemned the industry for indifference towards its popular image. This may be partly due, he felt, to 'a preponderance of technically

Advertisement for 'Cycolac' ABS. (Borg-Warner Chemicals, Europe)

This chair was made out of eighteen separate pieces.

The steel parts had to be pickled, ground and buffed.

Then welded together.

Then chromium-plated.

And the wood then had to be pressed into shape, then dried, then sanded and then varnished.

And then all the pieces had to be put together in sixteen more operations.

A lot of effort. And little inspiration.

This chair was moulded out of Cycolac ABS.

In one piece in a single ninety-second operation.

Little effort. And what a splendid inspiration.

I would like to know more about the many possibilities of Cycolac ABS Thermoplastics.

Please contact Mr. _____

job title _____

firm _____

address _____

Send to: Marbon UK Ltd.,
York House, Clarendon Ave.,
Royal Leamington Spa,
Warwickshire.

MARBON
DIVISION OF BORG-WARNER CORPORATION

BORG WARNER

CM 15/6

Cycolac ABS by Marbon
We'll help you turn to plastic.

Cycolac ABS applied by: Kartell, Overman, Furntrade, BMW, Citroën, DAF, Fiat, Ford, G.P.O. Lego, Remington, Samsonite, others.

oriented personnel in the industry who have tended to see technical advances and the mass marketing of the results as justified ends in themselves. The reactions of the final user have been somewhat overlooked.' And Mr Glanvill pleaded too for an effective system of plastics education.

The attitude of the public towards plastics has rarely been the subject of serious study, despite the industry's concern. Whenever a plastics design fails, the material itself is denigrated. The possiblity of misjudgement in other areas of the design process is rarely contemplated. A crazed mug, a cracked toy, a melted bowl are invariably considered material failures and not design failures. The ultimate form a moulding assumes is the result of a combination of many factors: the properties of the material itself, the limitations of the moulding process, the skill of the mould designer, the function of the article and definition of consumer needs, and finally the 'desire to make beautiful things'. As Alan Jarvis commented in *The Things We See* (Penguin Books, Harmondsworth 1946), it is unfortunately the last item, the superficial appearance, that actually sells the article, thereby making a mockery of the philosophy that good design is the outcome of a total design process.

For economic reasons, manufacturers usually employ plastics very close to their ultimate performance capability, and hence very close to failure-point. The more frequent the failures, the more the concept of plastics as cheap and inferior materials becomes reinforced.

Further problems have hit the plastics industry with the increase in oil prices. Petroleum took over from coal as the main source of raw materials for plastics in the early 1950s. Subsequently the industry expanded rapidly, and the years 1965-80 were scheduled to be an 'era of sophistication'. But the boom was interrupted by a general energy crisis in 1973, with shortages of all natural resources, followed in 1975 by a world-wide recession. A rethink had to take place. The speed of development had been too fast for resources to keep pace with demand. Between 1968 and 1973 alone, the consumption of the more popular thermoplastics doubled in the developed countries of the West. For too long quantity of production had been the paramount concern. A reaction set in, reflected by the present-day currency of the word 'ecology'.

As Dr James E. Guillet commented, we never actually consume, but convert from one chemical form to another. Since plastics are basically organic chemicals, they eventually rot and return to elements. Research into bio-degradable plastics, and into alternative sources to petroleum, has progressed. Most plastics are based on carbon, which is an endlessly renewable resource, and on hydrogen which might be derived from water. Pre-petrochemical days demonstrated how plastics can be adequately made from coal, limestone, air, salt and water, agricultural products such as grain, molasses, fats and milk, and of course wood and cotton. Leaving behind the 'Waste Age' we are now entering the era of alternatives, with a general return to re-usable containers and long-lasting products.

Plastics continue to develop. Whereas the previous generation had a very limited choice of moulding materials, the substances now at our disposal possess magical properties by comparison: materials lighter and more transparent than glass, stronger than steel, heat-resistant up to 2000°C; and bio-materials tolerated by the human body for replacing vital organs.

How different in the end are the molecules that control human genetic coding from those creating the synthetic materials that are now vital for supporting life on earth. Would the synthesis of life itself mark the beginning of the first truly 'Plastic Age'?

Appendixes

APPENDIX A   MOULDING PROCESSES

The techniques for shaping mouldable materials such as mud and clay are as old as man himself, and the centuries-old skills of casting metals, wax and plaster and of blowing glass have been adopted and modified for the new synthetic materials.

A basic knowledge of these principles is sufficient for understanding plastics moulding. The more complex aspects of the flow behaviour of resins and additives under heat and pressure are the particular concern of the mould-designer, a highly specialized artist in his own right. A final mould has to balance the designer's requirements against the chemist's formulae.

There are no fixed rules to determine which polymer should be used for a particular design. Many different plastics may be suitable, and several processes possible. The governing factor is the economics of the production run, and as the production increases it may be necessary to progress to other materials and other processes. Verner Panton's cantilevered chair is an example of such a progression. It was first hand-laid in GRP (FRP), then moulded in structural polyurethane foam, was later injection-moulded in nylon and was finally produced in ASA terpolymer.

Below is a list of the minimum production runs necessary to make the various processes economical:

Process	Production Run
Machining from stock mouldings	1–100
GRP (FRP) wet lay-up	1–300
Thermoforming	100–1,000
Rotational moulding	100–1,000
Cold matched-tool moulding	300–10,000
Foam moulding (expanded PU, PS)	min. 1,000
Extrusion	300–3,000 metres
Structural foam (PE, PP, ABS, PS)	1,000–10,000
Blow moulding	1,000–10,000
Hot matched-metal die moulding	10,000–100,000
Injection moulding	10,000–100,000

The above processes are also in order of increasing speed of production (an injection-moulded beer crate can be ejected in less than a minute), and similarly they are in rising order of the cost of tool making (a two-part injection mould for a chair shell machined from a solid block of steel can cost as much as a new house).

The earliest moulds for plastics were made of iron and steel, and the processes were highly labour-intensive. The development of chromium-plated moulds contributed much to the increased output of plastics in the thirties and hastened the era of mass-production. The plating eliminated the need to lubricate the mould, made it unnecessary to re-polish before each cycle, prevented corrosion, and produced an improved high-gloss finish on mouldings.

The introduction of beryllium copper moulds in Germany by Siemens and Halstre also cut down moulding time. They discovered in 1926 that the addition of a small amount of beryllium to copper produced a heat-treatable material, a 'temperable copper', and moulds made of this material can reproduce the finest detail possible on moulded plastics. Structural foam components for furniture are often moulded by beryllium copper moulds, reproducing the delicate characteristics of timber graining.

Moulds can be made in a wide variety of materials, including plastics themselves, depending on whether heat and pressure are involved. Steel and cast aluminium, GRP (FRP), unsaturated polyester filled with aluminium grit, epoxide, silicone and rubber are all commonly used.

Although most plastics are now synthetic materials, they can be conveniently machined by the same tools as those used for metals and hardwoods, the very materials they are in the process of replacing. The properties of plastics are closest to those of the softer metals, such as brass and copper, and interestingly suffer from the same build-up of internal stresses as metals. In plastics the process of orientation causes built-in stress, which is relieved by a heating process similar to annealing in metals.

Orientation involves stretching the resin under heat or at room temperature. It aligns the previously tangled molecular chains so that they flow in the same direction, producing a stronger material, since plastics are stronger along the direction of the molecular flow than across it, whereas timber is stronger against the grain. The process of blow moulding automatically orientates the molten plastic a certain amount. If polythene film is heated, the built-in stresses are released and the film shrinks, a property utilized in shrink-wrapping. Plastic hinges are a good example of orientation, where the molecular axis allows the hinge to be flexed millions of times, as on a polypropylene briefcase or box.

Polythene fibres, ski slopes, nylon brushes and carpets spring back into shape with what is called shape memory, the characteristic ability of plastics to return to their original moulded form. Synthetic shoes, for example, unlike leather, do not mould themselves to the wearer's foot shape, but return to their moulded form at the end of the day. When dentures were moulded from celluloid in the thirties, a blank was first moulded, then shaped and the teeth fitted. But under the effect of moisture, the denture would unfortunately revert back to its 'blank' shape, becoming a formless block.

Recent years have seen the publication of many books on craft approaches to plastics, but only a number of plastics can be worked easily by artists and designers in studios, workshops and colleges. The processes best suited to individual creativity are heat-forming thermoplastic sheets such as acrylic, polystyrene and polypropylene, twisting and carving acrylic rod and block, hand lay-up of GRP (FRP), casting expanded polystyrene, heat-welding polythene and PVC, and foaming soft and rigid polyurethane.

Compression Moulding

Compression-moulding is one of the oldest and least complex methods of moulding plastics to consistently close tolerances and has been used for over a century for rubber. Used mainly with thermosetting resins like phenol-formaldehyde, urea-formaldehyde and melamine-formaldehyde, it is a method which can make a wide variety of shapes. It is especially useful for large expanses and deep objects. If the articles are very small, a multi-impression mould is made which can mould many items at once, such as 'Bakelite' plugs and sockets or telephone hand-sets. Since the dies are cut from hardened steel, a multi-impression mould reduces expense.

The mould is loaded with a measured amount of resin in powder or pellet form, heated so that the resin flows, and then compressed by the male former until curing is complete. Although the process is simple, and inserts can be moulded in, cycle times are longer than those for injection moulding, and flash has to be removed from around the parting line of the moulding, thus increasing the cost. If thermoplastic resins are used, the mould must be cooled before the moulding is removed.

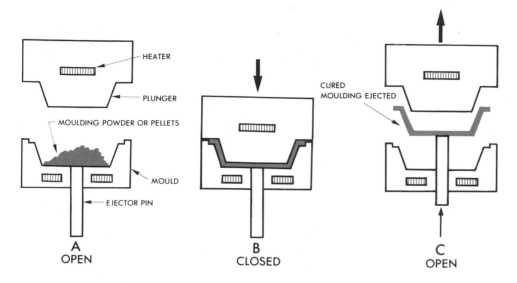

A
OPEN

B
CLOSED

C
OPEN

HEATER

PLUNGER

MOULDING POWDER OR PELLETS

MOULD

EJECTOR PIN

CURED
MOULDING EJECTED

COMPRESSION MOULDING

Transfer moulding

Sometimes in compression moulding the resin does not flow evenly around the mould. In 1926 an improved type of compression moulding was developed in which the pressure is not applied to the mould until the resin fills the cavity in a plastic state. Instead of a tool coming down directly onto the resin in the mould, the powder is held in a pre-heated chamber until it flows. Then the die descends, forcing the plastic resin out of the first chamber through a sprue into the mould below—the process of being 'transferred'. It was found that very thick mouldings produced in this way cured more evenly, and delicate inserts could be incorporated which would have been damaged by compression moulding.

Injection Moulding

Injection moulding has revolutionized the plastics industry more than any other process, rapidly churning out moulded objects which used to be cut individually by hand, such as combs from sheet plastic. It is one of the most versatile processes for moulding long-production runs of the same article. Initial costs are very high since the moulds must withstand high pressures, but it is an excellent method for mass-production.

Injection-moulding is normally used for the thermoplastic resins (polythene, polypropylene, polystyrene, nylon, PVC and ABS), but improved thermosets, such as polyester, have recently been developed as well as phenolics and aminos.

There are two basic processes and many variations combining features of both. The first process is similar to extrusion moulding except that it is not continuous but operates on a short cycle. In fact it was called extrusion moulding until 1938. The resin in granule or powder form is gravity-fed from a hopper into a heated cylinder where it is softened by a revolving screw and pushed forward. The screw then stops, reverses and then acts as a ram and forces the melt under high pressure into the cold mould. The moulding quickly cools and is ejected, usually in a completely finished state. The second process uses a plunger injection machine, inside which a hydraulic ram injects measured amounts of resin into the mould. Extremely high speeds can be achieved; four mould-openings a minute produces a very high output, and a polypropylene chair shell can be moulded in seventy seconds.

Moulds are usually two- or three-plate split moulds, the former being the most common and used for example to mould washing-up bowls, buckets, and open

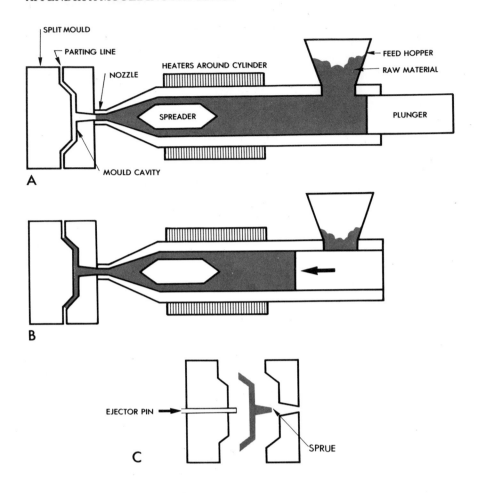

SPLIT MOULD

PARTING LINE

HEATERS AROUND CYLINDER

FEED HOPPER

RAW MATERIAL

NOZZLE

SPREADER

PLUNGER

MOULD CAVITY

A

B

EJECTOR PIN

SPRUE

C

PLUNGER INJECTION MOULDING MACHINE

(SCREW TYPE MACHINES HAVE AN ARCHIMEDEAN SCREW INSIDE INSTEAD OF A PLUNGER)

A 'short-shot' of injection-moulded combs with sprue and runners still attached. This occurs when there is insufficient resin to fill out all the mould cavities. (Photo by courtesy of Ronald Beck)

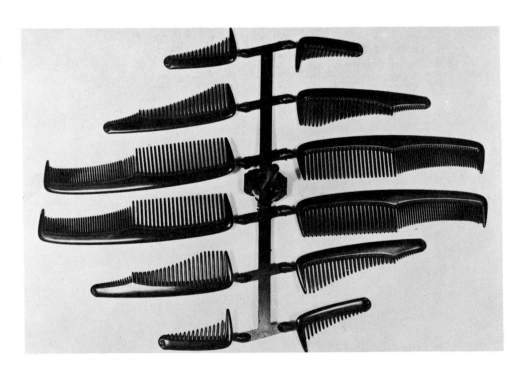

containers of all kinds. Hollow articles such as bottles are blow- or rotational-moulded.

The two parts of the mould meet along the parting line, which can be identified on injection-moulded articles by a raised line of flash. Occasionally the parting line is disguised along the rim of an article or some other detail such as a groove. A circular mark, sometimes sharp, indicates where the sprue was broken off.

As in compression moulding, many small articles can be produced in a multi-impression mould, with each moulding fed from a central gate. Golf tees used to be moulded in the form of spokes on a wheel and later snapped off. The illustration shows what is termed a 'short shot'. The resin has not filled out the mould cavity. A multi-impression mould must be perfectly balanced, with each cavity an equal distance from the gate. All injection mouldings are slightly tapered, and never totally right-angled, so that they can be ejected from the mould, and there are many other aspects to consider, such as shrinkage of the moulding, and sink and flow marks.

Extrusion and Pultrusion

Extrusion is a process for producing long continuous mouldings of uniform section, from rods, tubes, filaments, sheet and film to more complex sections, such as curtain rails.

Dry granules or powder are fed from a hopper into the machine and spiralled through a heated die by a continuously revolving Archimedes screw. As the emerging extrudate is hot, it will lose its shape unless it is cooled quickly by water or jets of air.

The affinity of this process with the making of pasta is more than visual. The principle of the pasta-moulding machine was adapted for use with the early plastics such as rubber and shellac. But the first thermoplastic to be forced through such a machine was celluloid, by the Hyatt Brothers in 1875-80. It was not until the 1930s that plastics were extruded commercially. A recent development is the co-extrusion of different materials such as metal with PVC or metallized ABS for car accessories. The packaging industry in particular uses these laminates, often combining the properties of a cheap and an expensive material.

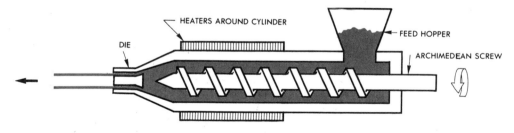

EXTRUSION

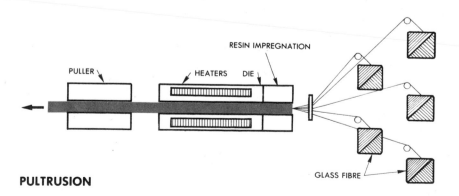

PULTRUSION

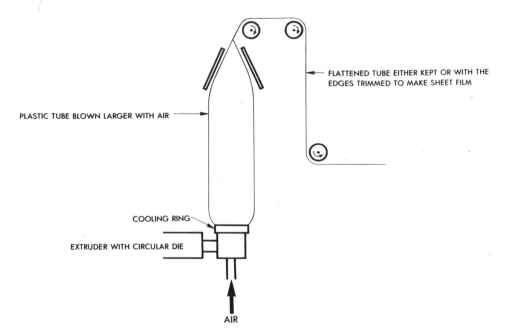

FLATTENED TUBE EITHER KEPT OR WITH THE EDGES TRIMMED TO MAKE SHEET FILM

PLASTIC TUBE BLOWN LARGER WITH AIR

COOLING RING

EXTRUDER WITH CIRCULAR DIE

AIR

EXTRUSION OF LAYFLAT FILM AND TUBING

In the early days of casein and celluloid, only small-scale items such as knitting needles were extruded, but now extremely large extrusions are possible, such as high-density polythene piping 150cm in diameter (almost five feet). Flat film is extruded through a slit die and wound over cooling rollers. Alternatively it is known as layflat tubing and is made by blowing, which is now the standard process. A tube of hot resin is extruded through a circular die, blown out into a larger tube and flattened over rollers. The edges are then cut, producing a double layer of film.

Pultrusion is one of the latest moulding techniques, and is so new that design parameters are not yet established. It is the opposite to extrusion, and, as the name implies, the resin is pulled through a die instead of pushed. Like extrusion it is a continuous process, producing all shapes and sizes of rods, angles and sections.

During the process, the reinforcing fibres or glass matting are pulled through a long die and held under tension so that the fibres are tightly aligned and not twisted. This process produces a moulding which is extremely strong longitudinally. Inside the die the fibres are impregnated with resin and cured as they pass through.

To date, the process has only been used with reinforced resin—originally resin reinforced with carbon fibres, and later glass-reinforced polyester. Like extrusions, pultrusions can be moulded around a central core.

Blow moulding

The blow moulding process is used for the manufacture of hollow objects such as balls, toys and bottles from thermoplastic resins (polythene, PVC and polypropylene) and is based on the principle of glass blowing.

A parison (hollow tube) of heated and softened material is extruded or injection-moulded into a two-part female mould. The mould is closed and air is injected through a hollow blowing pin, forcing the tube to expand against the inside walls of the mould, where it cools and is then removed.

The earliest blow-moulded plastics articles were table-tennis balls, and dolls' heads and bodies, blown from celluloid. They were made by placing two thin sheets of celluloid between pre-heated hollow moulded plates, with air injected through a small nozzle pushed between the sheets to make the soft plastic inflate, and causing the two sheets to weld together. The mould was cooled by dipping in water before removing the moulding. After World War II the availability of polythene led to the introduction of blow moulding from tube.

The greatest problem with blow moulding is to ensure a correct wall thickness at particular points of stress, such as corners and bottle necks. Now automatic programming can determine this in advance by varying the thickness of the extruded parison walls. As high pressure is not involved, it is a relatively cheap process.

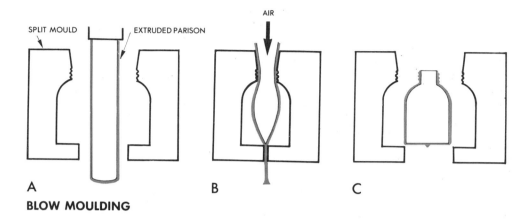

BLOW MOULDING

Rotational moulding (Also called rotational casting and rotomoulding).

Originally adapted from the slush casting of metals, and similar in principle to the modern slush casting of ceramics and chocolate Easter eggs, this process, like blow moulding, is used for moulding hollow objects but can reach a much larger scale. Great quantities of footballs and dolls are made in this way, and all types of domestic and industrial containers from TV cabinets, luggage and furniture to dustbins and storage tanks.

The process uses thermoplastic resins, particularly high- and low-density polythene, PVC, polystyrene and nylon. The powdered resin is placed inside a cold mould, heat-softened inside an oven, and rotated gently about two axes, tumbling the resin so that it coats the inside of the mould evenly, like a skin. It is cooled as it rotates and the result is a seamless hollow object with no built-in stresses, since no high pressures have been built up inside. The moulds are therefore relatively cheap; also, one machine can hold several small moulds at once, and inserts can be included in the moulding.

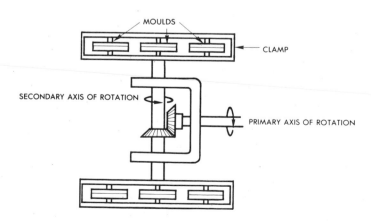

ROTATIONAL MOULDING

Modern methods of rotational moulding were developed in 1958 when low-density polythene became available in a suitable form. Spheres have been moulded 5 metres (16½ft) in diameter, and containers over 10 metres (33ft) have been made for military use in America. More recently the casting of layers of different resins, such as polystyrene backed with high-density polythene for increased strength, has widened the possibilities of the process.

Slush moulding

Slush moulding is a simpler type of rotational moulding for use with PVC plastisol (PVC powder in a plasticizer). A great many 'squeaky' toys are made in this way. The plastisol is poured into a heated mould, rotated, and the excess resin is poured off. After curing in an oven, the mould is opened and the moulding removed. Low pressures mean that cheap moulds can be made from metal alloy.

Calendering

Calendering was originally a paper and fabric process, and in the 1930s it was the standard process for applying rubber to fabric in the manufacture of waterproof clothes and curtains. Today sheet and film can be produced by calendering as an alternative to extrusion. Hot thermoplastic resin is squeezed through a series of heavy steel rollers rotating in opposite directions, rather like laundry in the pre-spin-dryer era. The chief material processed in this way is vinyl, making film, sheet and flooring of all types. Embossed rollers can be used to apply a texture and patterns can be easily printed. Several layers of sheet can be laminated together.

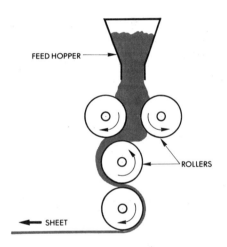

FEED HOPPER

ROLLERS

SHEET

CALENDERING

In the 1890s celluloid film was cast on sheets of glass and used as a carrier for the gelatine emulsion of early photographic film. Another process was wheel casting in which a purified solution of resin such as celluloid or cellulose acetate was poured onto a revolving drum. The solvent was gradually evaporated out of the resin by heat as it rotated and the dried film was finally stripped off and hung up to 'season'.

Dip moulding

This is a process now almost entirely confined to vinyl (PVC plastisol), and rubber, although cellulose acetate was moulded into medical containers in this way in the thirties. Dip moulding produces cheap flexible components such as sleeves, casings, cable shrouds, bellows, pipe connectors, ducting and hand-grips for tools. A metal former, usually aluminium, is machined to the required internal dimensions of the moulding. It is then heated and dipped into the viscous polymer, which forms a skin over it. A cold former picks up more detail although the coating is thin. Wall thickness depends on the length of time the former is immersed in the resin. It is then taken out, cured in an oven, and when cool the flexible seamless moulding is stripped off the former.

Casting and Embedding

Based on the ancient techniques of casting metals in open or split moulds, casting is one of the simplest of all moulding techniques, since no pressure is required. Resins that become liquid during polymerization can be cast, the main ones being acrylic, polyester and epoxide and, formerly, phenolic. The process is used for encapsulating electronic components, casting dental plates in acrylic, and for embedding natural specimens, as well as many decorative uses.

Leo Baekeland announced the first casting resin in 1907, phenol-formaldehyde. The resin was rough-cast by pouring it into cheap lead moulds, either straight moulds or split moulds (two halves clamped together to allow undercuts). Some cast resins cure without heat, but even in ovens at moderate temperatures the early

PVC dip mouldings for the electrical and automotive industries by Plastic Dip Mouldings Ltd. (Photo Chris Smith)

177

phenolic castings took as long as three to six days. The casting was finally punched out, sanded and polished.

Phenol-and urea-formaldehyde cast tubes, rods, bars and sheet were sliced up and carved by grinding and shaping into decorative items such as jewelry and candlesticks. Buttons were cut from rod, and shaped on semi-automatic lathes to give them a bevelled edge. Stone cutting blades applied details such as slots and grooves, and the thread holes were drilled. After a final polish the buttons were attached to cards.

In 1947-8 cold-setting polyester resins were developed, and it became possible to simulate what natural fossil resins such as amber had achieved fifty million years ago—the embedding of insects and flowers. Building up layer by layer, it is possible to position the encapsulated objects in any arrangement.

Finishing and Decorating

Integral Moulded Effects

1. Leather graining became very popular in the sixties, although cellulose nitrate and PVC leathercloth had been made in the early days of the industry. Polypropylene, high-density polythene and ABS accept a grained texture easily. The textures of leather, timber, fabric or weaving are etched into the surface of the mould. More recently a high degree of realism for timber graining has been achieved by high-impact polystyrene structural foam.

2. Printed designs are incorporated into the top surface of decorative laminates, such as 'Formica', and labels and foils are placed inside the mould and fused onto children's melamine tableware.

3. Mixed colours, which are more the result of incomplete blending, can produce very attractive effects. In the twenties and thirties particles of other materials were used to create shimmering designs.

Post-moulding Surface Treatments

1. Printing processes are used, such as *silkscreen* and *lithography,* as on yoghurt cartons and kitchen detergent bottles. Before the development of satisfactory colour mixing, articles had to be sprayed or even hand-painted. These finishes were often applied to the reverse sides of transparent mouldings to conceal dark glue marks.

2. The hot stamp process is a common method for printing the gold and silver labels and trade-marks on cosmetic and other containers. A heated die in the shape of the embossed design is pressed into the surface of the moulding through a foil.

3. Metal plating on plastics is being increasingly used as tough new thermoplastics replace metal components, particularly on cars. The two methods used are vacuum metallizing and chromium electroplating. In the first process a thin coating of metal is deposited on the primed plastic surface in a vacuum chamber. The air is evacuated and an activated current causes the metal to evaporate and form a vapour which coats the component. A final coating of lacquer is applied.

The second process produces a better bond, since the coating is deeper. The plastics moulding is first etched, or sensitized, then plated with nickel or copper, and finally chromium-plated. Almost all thermoplastics can be chromium-plated, usually with copper or silver.

4. Metal inlays, once common in America in the late thirties and early forties, were used for decorative designs. A flat metal design was laid at the bottom of a compression mould before the powder was added, and thus it became incorporated in the moulding. This technique was used to make decorated 'Bakelite' box lids, for example. Another method akin to the standard techniques of inlaying timber involved placing strips of a hardened metal into grooves in the moulded object, with the metal inlay laid on top. Applied pressure forced the lower strip to expand into the plastics.

At the Plastics and Rubber Institute, London, some fine examples of Parkesine can be seen, the forerunner of all modern plastics. Moulded as early as the 1850s and 1860s, they are inlaid with mother-of-pearl and metals.

Plastic Coatings

Metal, paper and fabric can be coated with plastics using a variety of techniques, such as dipping, fluidized-bed coating, electrostatic spraying, brushing, roller-and-knife coating and curtain-coating.

Fluidized-bed Coating

This process was developed by BASF in Germany. A heated metal component such as a kitchen drainer or chair frame is suspended inside a tank containing a bed of fine, dry powdered polymer. Compressed air is blown through the box causing the plastics powder to stick to the metal, which is fused on a second heating.

Electrostatic Spraying

Developed in the 1930s this process produces fuzzy-coated articles such as riding hats, paper and fabrics. The article to be flocked is first coated with adhesive, earthed, and electrically charged plastics particles are sprayed on to it. Plastics are good electrical insulators and when so charged the particles repel each other ending up parallel and at right angles to the base surface, even if it is curved.

Thermoforming

Thermoforming is strictly a fabricating rather than a converting process since it involves ready-moulded plastics in the form of sheet, film, rod or tube. Thermoforming is a general term covering the various processes which shape plastics by softening with heat, thus using only thermoplastic materials such as acrylic, polystyrene and polypropylene.

Typical thermoformed mouldings are illuminated signs, lighting units, bathroom wall-units, packaging, yoghurt pots, and inner trays for chocolate boxes. Acrylic baths are thermoformed from sheet, together with ABS boat hulls, cosmetic and camera point-of-sale displays, and polystyrene disposable cups by the million.

Moulds can be made cheaply from plaster and timber, or for longer runs, metal and filled epoxide resin. There are three main methods for thermoforming sheet, which can themselves be combined.

Drape forming (or slip forming)

Drape forming is simply the draping of a heat-softened sheet over a mould. Many acrylic aircraft parts such as cockpit covers and bomber noses and windows were made this way during World War II; they were laid over cloth-covered wooden formers and clamped there until cool. Early heating methods included baths of mineral oil, steam tables, electric plates and infra-red lamps. Automatic machines now accomplish the whole process with pneumatic rams which raise the male mould, forcing it into the softened sheet; a vacuum sucks the sheet over the mould.

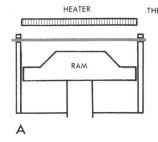 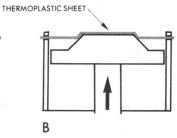 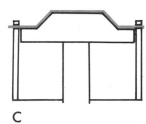

A   B   C

DRAPE FORMING

Vacuum forming (or female forming)

Vacuum forming is one of the simplest and most widespread methods for forming shapes which are not too deep. The thermoplastic sheet is clamped over the mould, heated and then vacuum-sucked down into the mould. A plunger or plug can be used to push the sheet into deeper cavities.

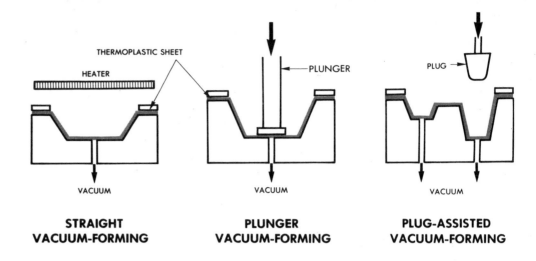

STRAIGHT VACUUM-FORMING

PLUNGER VACUUM-FORMING

PLUG-ASSISTED VACUUM-FORMING

Bubble forming (or air slip)
Bubble forming is basically drape forming with the sheet pre-stretched by air, which reduces thinning of the material at corners. Free blowing does not need any mould at all, and is used to shape domes and skylights.

Blister packs, now a familiar method of packaging small articles such as toys and loose curtain hooks and drawing pins, are made by thermoforming plastic sheets over the article by vacuum, onto a card backing.

Shrink wrapping is a packaging method for wrapping heavy, bulky multipacks of beer cans, wine bottles, jam-jars, etc., in boxes or on pallets. The load is wrapped in plastic film such as polythene, PVC or polypropylene, and passes through a shrink tunnel. The application of heat causes the film to shrink tightly around the package.

Stretch wrapping is similar to shrink wrapping but no oven or heat tunnel is necessary. It is mainly used for wrapping food such as meat, cheese and fruit in polystyrene or pressed fibre trays in supermarkets. The development of suitable stretchable self-sealing film is considered to be as important a breakthrough as the polythene bag, for cellulose film simplified the hygienic display of fresh food without condensation.

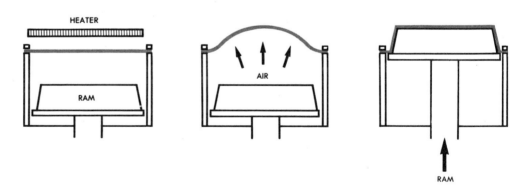

BUBBLE FORMING

Moulding of Fibre-Reinforced Plastics (GRP, FRP)
Thermosetting resins are almost always reinforced in order to reduce brittleness and increase strength. Phenol-, urea- and melamine-formaldehyde are reinforced with fillers such as wood flour, mica, fabric and paper. A few thermoplastic resins (such as nylon, polypropylene and acrylic) are reinforced with glass fibres and processed on injection-moulding equipment. But usually the term 'fibre-reinforced plastics'

refers to glass-reinforced plastics, notably polyester and epoxide resins. The section on glass-fibre-reinforced plastics (p. 130) describes nine processes for moulding these fibre-reinforced composite materials.

Lamination

The first decorative laminates were made by the Formica Insulation Company, a Westinghouse subsidiary in the thirties. Lamination is a reinforcing process whereby strong composite boards are hot-pressed together in a way similar to plywood veneering. Sheets of paper or fabric, usually kraft paper, are impregnated with a phenolic thermosetting resin. Since it is a more expensive resin, only the top sheet carrying the pattern and the final overlay of alpha-cellulose are impregnated with melamine. The alpha-cellulose turns transparent during the pressing, to give an extra scratch-proof, heat- and chemical-resistant surface. The pads of impregnated paper are then compressed between highly polished steel plates in a heavy daylight press, and cured at a temperature of around 302°F. Many such pads are compressed at the same time. A sanding machine is used to give the board a matt surface if required, or a texture can be embossed on the surface during pressing.

Many other thermosetting resins may be used for lamination, such as urea-formaldehyde, resorcinol-phenolic and neoprene, depending on the function of the laminate. This process is now used to supply cheap construction materials such as chipboard and hardboard, which can be surfaced with attractive plastic decorative laminates.

APPENDIX B GLOSSARY

(For terms not included here, see the explanations given in the text.)

Accelerator An additive which speeds up the curing of a synthetic resin by acting on the catalyst. Similar to a hardener.

Alpha cellulose Very pure cellulose treated with chemicals.

Aminoplasts (Aminoplastic resins) Resins made from amino compounds, i.e. compounds derived from ammonia. Mainly urea-formaldyhyde and melamine-formaldehyde.

Annealing The heating of a plastic material to allow the stresses built up during moulding to be relieved without distortion.

Anti-oxidant Counteracts the effects of degradation by oxygen or ozone on a plastic material during manufacture and use.

Anti-static agent Reduces or eliminates electricity in plastics.

Atactic Irregular spatial arrangement of molecular groups around a main polymer chain.

Autoclave A steel cylinder for carrying out reactions under heat and pressure, for low-pressure lamination (see Rubber-bag moulding) and for sterilizing.

Benzene ring The most important aromatic hydrocarbon, consisting of a ring-shaped molecule of six carbon atoms and six hydrogen atoms.

Biaxially oriented film Plastic film stretched both longitudinally and laterally during processing. It is extremely tough as the molecules are aligned along the direction of the stretch (see Orientation).

Blowing agent An additive that produces expanded or foamed plastics. It is either added as a chemical into the mix, producing inert gas bubbles on heating, or is blown as a gas into the melt before it cures.

Carbon fibres Very strong fibres made by the carbonization of oxidized polyacrylonitrile.

Catalyst (see also Accelerator and Hardener) A substance taking part in a chemical reaction to initiate or speed up polymerization without being used up itself.

Cellular plastics Expanded or foamed plastics with a cellular, sponge-like structure. Expanded plastics have a uniform structure of non-intercommunicating cells: foamed plastics are porous with intercommunicating cells.

Cold Moulding Materials such as gutta percha, certain phenolic resins and inorganic compounds like asbestos cement and bitumen can be shaped by pressure at room temperature. The mouldings can be removed and cured in ovens if necessary.

Cold-curing plastics Polymers which cure at normal temperatures without applied heat.

Collodion Cellulose nitrate soluble in alcohol or ether.

Compounding Mixing the various additives into the basic resin to produce a compound.

Copolymer A plastic material produced by the polymerization of two or more chemically different monomers.

Crimping Giving a curl to synthetic fibres.

Cross-linking A process in which the molecular chains of a resin are linked together with chemical bonds, producing a material which cannot be re-softened and re-moulded. Similar to Vulcanization, q.v., of rubber.

Curing The completion of the chemical reaction in moulding, usually implying hardening, produced by either additives such as catalysts or by applied heat or pressure.

Cycle The moulding cycle is the time taken to complete the repeating sequence of operations on a moulding machine.

Die Moulds are often called dies, although the word refers specifically to the steel hole through which the resin is forced out of an extruder.

Dielectric Possessing good electrical insulating properties. All plastics are electrical insulators.

Dielectric heating (Also called electronic, radio or high frequency heating) A method of pre-heating plastics materials by high-frequency currents; used, for example, to weld sheets together.

Dip coating Dipping an article into either viscous resin or air-suspended powdered resin (fluidized) to obtain a certain thickness of coating.

Elastomer A rubber-like synthetic plastic, which when stretched at room temperature will return to its original shape.

Electrostatic spraying A process for coating an object with electrically charged particles.

Encapsulation (Potting) Embedding an electrical component or zoological specimen in clear casting resin.

Erosion casting Moulding casts from humans or animals by injecting selected areas (e.g. lungs) with polyester resin and eroding away the remaining tissue by dipping in acid.

Fabricating Making finished articles by processes other than moulding. Ready-moulded sheet, rod and tubes are machined, glued, thermoformed, etc.

Feedstocks The raw materials of the plastics industry, e.g. ethylene, benzene, butadiene, derived from petroleum, natural gas or coal.

Fibre Single filament or thread produced by forcing molten plastic through a Spinneret, q.v.

Filament winding A process for moulding tubes by winding resin-impregnated fibres onto a mandrel (a cylinder or rod).

Filler Inert material added to a compound either to improve its properties, or to reduce costs by extending it. Common fillers are wood flour, talc, asbestos, cotton flock and carbon black. Glass fibres are added to plastics such as polyester, polymethyl methacrylate, nylon and polypropylene.

Flash Resin which has been forced out of a mould along the Parting lines, q.v., during moulding. It has to be removed, but can leave a slightly raised line.

Flow mark A defect on the surface of a moulding caused by inadequate mixing. The flowing plastic partially hardens into a line.

Fluidized-bed coating A process for coating an article by heating and dipping into a tank of dry thermoplastic powder, which is made to float and adhere to the article by the injection of compressed air.

Former A shaped mould used for dip-coating plastics, as used in rubber gloves and boots.

Gate The narrow inlet in injection and transfer moulding through which the plastic is injected from the runners into the mould cavity. It allows the Sprue and Runner, qq.v., to be easily broken off the finished moulding.

Hardener Promotes and controls the curing (hardening) of a polymer.

Heat welding A heat and pressure process using high-frequency welding, spot and seam welding or jets of hot air or gas, for bonding plastics to itself or other materials.

High polymers A word synonymous with plastics and super-polymers. Materials composed of giant molecules with molecular weights in the thousands.

Hydrolysis The chemical decomposition of a compound through the addition of water.

Hydroxyl groups (OH) Very reactive groups of molecules consisting of atoms of oxygen linked to atoms of hydrogen, used to build many plastics.

Hygroscopic Tending to absorb water. Plastics such as casein, urea and nylon can warp, crack or change dimensions when affected by water.

Impression The cavity inside a mould which gives the shape to the moulding. Multi-impression moulds have many connected cavities.

Isomeric polymers Plastics composed of molecules which are similar but which are arranged in different ways to produce different properties.

Isoprene (C_5H_8) A colourless liquid monomer composed of molecules of carbon and hydrogen. Easily compounded into rubbers. Gutta percha and natural latex rubber are composed of long chains of isoprene molecules.

Isotactic A particular molecular structure in which groups are placed on one side of the main carbon chain, forming a regular pattern. (See Stereospecific plastics)

'K' Value The measurement of the viscosity of a plastic substance when dissolved in a solvent.

Kraft paper Paper made from sulphate wood pulp, used in the manufacture of decorative laminates.

Layflat (or blown) film Film produced as an extruded blown tube, cooled, then flattened and transferred to rollers.

Leathercloth Upholstery material made of fabric coated with plastics such as cellulose nitrate or PVC, given a grained or embossed texture.

Masticating machine A machine invented in 1820 by Thomas Hancock for making crude rubber into a mouldable dough by kneading and heating it with spiked rollers.

Matrix A lattice of resin and reinforcement bound together to form one mass.

Melt Thermoplastic resin in a molten state ready to be moulded.

Melt Flow Index (M.F.I.) The quantity of a thermoplastic resin that can be forced through a specially sized opening in ten minutes at $190°$C under constant pressure.

Methane (Marsh gas) A colourless gas composed of the simplest organic molecule, CH_4, one carbon atom with four hydrogen atoms attached to it.

Mica A group of phyllosilicate minerals used as fillers in resins such as phenolic and shellac for their insulating and low water absorption properties.

Molecule The smallest particle of a substance capable of separate existence, formed by the linking of two or more atoms.

Molecular weight The sum of the atomic weights of all the atoms in a molecule.

Monofilament Continuous thread, usually extruded and thicker than textile fibres, cut into lengths and used for bristles, fishing line and rope, etc.

Monomer The simple, low-molecular weight compound which through repetition caused by polymerization forms the chain structure of a High polymer, *q.v.*

Moulds The tool for shaping plastics under heat and pressure, usually consisting of a hollow die (female mould) and a shaped plunger (male mould). Moulds may be split into several movable parts for complex mouldings and undercuts, or have many cavities within the same mould.

Multifilament A continuous thread made up from monofilaments bound together.

Naphtha An inflammable, volatile liquid derived from the distillation of petroleum or coal.

Neoprene A synthetic rubber copolymer developed in 1932 by Du Pont. Now a generic term for copolymers made from chloroprene.

Olefins A family of unsaturated active hydrocarbons with one carbon-to-carbon double bond. The olefins are named after their corresponding paraffins by the addition of '-lene' or '-ylene', e.g. ethylene and propylene.

Organic compound A compound with carbon in its molecular structure. Plastics are made mainly from organic compounds.

Orientation When a plastic resin is stretched during moulding the molecular chains align themselves in the direction of the pull, strengthening the material along that line. Many plastics are oriented during processing, e.g. films and fibres. Orientation produces built-in stresses which can be relieved by heating.

Parison An extruded tube of plastic which is moulded into hollow articles by blow-moulding or injection blow-moulding.

Parting line The line along which the two moving parts of a mould meet, identified usually by a raised line of flash on finished mouldings.

Pattern The solid form, usually of wood, from which the mould is made. Chisel marks and graining can be found on structural foam mouldings which simulate timber.

Plasticizer A chemical substance added to a compound to increase its flexibility and softness and to improve its flow properties in the mould.

Polymer A chemical compound made of large molecules formed by the linking together of many smaller repeating molecular units (monomers) through the process of polymerization. The word polymer is synonymous with plastic material. (See also Copolymer and Terpolymer).

Polymerization The process of forming a new, high-molecular compound by joining up smaller units (monomers) into a chain.

Polyol An alcohol (a certain type of hydrocarbon) which has many hydroxyl groups, e.g. polyester and polyether are polyols used to make polyurethane foam.

Preform Compressed plastic tablets or pellets of a certain weight for placing in a mould instead of powder. They are easier to handle, reduce the cycle time and, if pre-heated, flow better.

Profile cutting In polyurethane foam this is the 'egg-box' effect resulting from passing a slab of foam between rollers spiked with hammers, which compress the sheet as it is cut with a knife.

Reinforced plastics Plastics strengthened with various fillers and moulded at low pressures. The term often refers in particular to glass-fibre or asbestos reinforced polyester, or epoxide laminated mouldings.

Resin A term originally referring to natural resins but now mainly indicating synthetic compounds.

Rubber-bag moulding A moulding process in which a rubber membrane is sucked down by air or fluid pressure over material laid over a hard mould. It is used for laminating reinforced plastics or for forming curved plywood.

Runner In injection moulding the runner is the groove which carries the molten resin from the Sprue to the Gate, *qq.v.*

Saturated compound A compound in which there are no free links available since all the atoms are attached to other atoms.

Shrink wrapping A process for packaging articles by heating plastic film to release the in-built stresses, causing it to shrink around the object.

Sink mark A small depression on the surface of a moulding caused by contraction of the material, particularly if there is a rib on the other side.

Sintering A process in which fusible powder is solidified without melting, by compression and heat. The term is also used for coating articles which are first heated then plunged into plastics powder which fuses to the surface.

Slush moulding A simple process for moulding hollow shapes, by pouring resin into a heated mould and pouring off the excess, leaving a 'skin' to cure.

Solvent Usually a volatile liquid in which a polymer is

dissolved to form a solution. Important for adhesives and coatings.

Spinneret A die with sometimes thousands of small holes through which plastic is extruded or drawn in the manufacture of fibres and filaments.

Split mould A mould consisting of two or more movable parts (splits) capable of moulding undercuts and shapes not possible with straight male and female moulds.

Sprue The passage in injection moulding through which the plastic flows from the nozzle of the injection machine into the runners and thence into the mould cavity. The term also refers to the hardened material in that passage which is broken off after the moulding is ejected.

Stabilizer A substance added to plastics to prevent deterioration during processing and use.

Staples Short fibres used for spinning textiles, often blended with other fibres.

Stereospecific plastics Plastics with their molecules arranged in a pre-determined structure with side chains oriented in the same direction. (See Isotactic).

Structural foam Rigid plastic with a strong skin over a foamed core. A characteristic swirl pattern can usually be seen on the surface.

Syndiotactic A molecular structure in which groups of molecules are placed on alternate sides of the main chain.

Terpolymer A plastic produced by the polymerization of three different monomers.

Thermoforming A general term describing the shaping of thermoplastic sheet with heat and/or vacuum.

Thermoplastic A plastic material capable of being re-softened and re-shaped with heat.

Transfer moulding A process for moulding thermosets in which the material is first heated in a chamber, and then transferred to a closed heated mould to cure.

Ultrasonic bonding A method of welding plastics using ultrasonic waves transmitted through a metal tool, causing the plastic surfaces to fuse together. It is used where the materials are not easily bonded by heat welding.

Vacuum metallizing A method of coating plastics with metal in the form of vapour.

Vulcanization The chemical reaction in which rubber is cross-linked by sulphur or other substances, making it elastic.

APPENDIX C
MAJOR EXHIBITIONS, COMPETITIONS, AND HISTORICAL FILMS

Some of the exhibitions and competitions listed below are entirely devoted to plastics, others contain certain plastics designs of interest.

1851 The Great International Exhibition, Crystal Palace, London. Display of gutta percha and hard rubber.

1862 The Great International Exhibition (also called the Universal Exhibition), London. Parkesine first exhibited.

1925 The Wembley Exhibition. First display of 'Beetle' thiourea-formaldehyde.

1935 The First Annual Plastics Exhibition, sponsored by Modern Plastics in New York.

1935 British Art in Industry Exhibition, Royal Academy, London. 'Specimen Room' displaying plastic mouldings.

1936 First Modern Plastics Competition (sponsored by Modern Plastics).

1937 Art in Industry Exhibition, Paris. Plastic school furniture.

1938 Paris International Exhibition. The Blue Ribbon Tower in cellulose acetate.

1938 The 'Bakelite Travelcade' travels the USA with its film *The Fourth Kingdom*.

1939 'L'Age des Plastiques' Exhibition, Paris.

1940 'Organic Design in Home Furnishings' Competition, Museum of Modern Art, New York. First plywood chair shells formed with synthetic gluc by Eames and Saarinen.

1941-53 The John Wesley Hyatt Award organized by the Hercules Powder Co., USA.

1946 'Britain Can Make It', Council of Industrial Design Exhibition at the Victoria and Albert Museum, London. Mouldings resulting from war technology.

1949 International Low-Cost Furniture Competition, Museum of Modern Art, New York. First plastic chair shell in GRP by Charles Eames.

1951 The British Plastics Exhibition, the first plastics industry exhibition in Europe.

1957 The Bachner Award founded by the Chicago Moulded Products Corp., USA.

1957 Alain Renais directs *Le Chant du Styrène*, a short industrial film for La Société Péchiney. Wins a Venice Golden Mercury Award the following year.

1960 'Design in Plastics' Competition, sponsored by Shell.

1961 Dunlop Ltd establish the Aeropreen Award, now called the Dunlopillo Designer Award.

1964 'Polypropylene Design' competition sponsored by Shell.

1966 'Plástica con Plásticos' Exhibition, Buenos Aires.

1967 'Prospex 67', at the Royal College of Art, London.

1968 'Structures Gonflables', Musée d'Art Moderne, Paris.

1968-9 'Plastic as Plastic' Exhibition, Museum of Contemporary Crafts, New York.

1969 'Research in Plastics Design', Palazzo Reale, Milan.

1971 'Designing in Plastics' Exhibition, Design Centre, London.

1972 'Design and Plastics', Museum of Decorative Arts, Prague.

1972 'Italy: The New Domestic Landscape', Museum of Modern Art, New York.

1975 'The Chair in Plastics' Exhibition, Centrokappa, Milan.

1975 'New Materials', Society of Industrial Artists and Designers, Science Museum, London.

1976 The first Prince Philip Award announced by the Plastics and Rubber Institute, London.

1977 'Plastics Antiques' exhibition organized by Wolverhampton Polytechnic and BIP Ltd.

1977 An historical and educational plastics exhibition, (scheduled) Smithsonian Institution, Washington.

The principal current plastics trade exhibitions:
Interplas, Great Britain
Kunststoffe, West Germany
Plast, Italy
Europlastique, France
The National Plastics Exposition, USA.

APPENDIX D
CHRONOLOGY OF THE FOUNDING OF ORGANIZATIONS AND PERIODICALS

1925 *Plastics* founded in America, the first plastics magazine; became *Plastics and Molded Products,* then in 1963 *Modern Plastics*.

1929 *British Plastics* published, London

1929 British Plastics Federation founded, London.

1930 Society of Industrial Artists and Designers founded in London, the first society of industrial designers.

1931 The Institute of the Plastics Industry founded, London. In 1947 became the Plastics Institute, since 1975 called the Plastics and Rubber Institute.

1937 Society of the Plastics Industry founded, New York.

1941 Society of Plastics Engineers established, Greenwich, Connecticut

1941 The Plastics Industries Technical Institute opened in Los Angeles, California, the first technical college for plastics.

1944 Council of Industrial Design set up in London under the Coalition government.

1961 Plastics Institute of America founded, New Jersey.

1975 The Plastics Institute merges with the Institute of the Rubber Industry.

1975 The Science Research Council (SRC) announces the setting up of a Polymer Engineering Centre.

Bibliography

Books

Ambasz, Emilio (ed.) *Italy, The New Domestic Landscape*, Museum of Modern Art, New York, 1972

American Hard Rubber Co., *Hard Rubber and Plastics Handbook*, New York, 1948

Arnold, Lionel K., *Introduction to Plastics*, Allen and Unwin, London, 1969; Iowa State University Press, 1968

Barthes, Roland, *Mythologies*, Paladin, London, 1973

B.A.S.F., *In the Realm of Chemistry*, Badische Anilin- und Soda-Fabrik, Ludwigshafen-am-Rhein, 1965

Banham, Reyner, *Theory and Design in the First Machine Age*, Architectural Press, London, 1960

Beck, R. D., *Plastic Product Design*, Van Nostrand, New York, 1970

Boenig, H.V., *Polyolefins: Structure and Properties*, Elsevier, Holland, 1966

Braley, Silas A., *Biomaterials in Clinical Use*, Dow Corning Center for Aid to Medical Research, Michigan, 1975

The Chemistry and Properties of Medical Grade Silicones, Dow Corning Center, Michigan, 1970

The Use of Silicones in Plastic Surgery, Dow Corning Center, Michigan, 1972

Acceptable Plastic Implants (reprinted from *Modern Trends in Biomechanics*), Butterworth, London, 1970

Brydson, J.A., *Plastics Materials*, Illiffe, London, 1966

Buttrey, D.N., *Plastics in Furniture*, Applied Science Publishers, London, 1976

Plastics in the Furniture Industry, MacDonald, London, 1964

Carr, A.F., *A Model Course in Plastics for Students of Industrial Design*, The Plastics and Rubber Institute, London, n.d.

Cook, J. Gordon., *Your Guide to Plastics*, Merrow, Watford, 1964

Couzens, E.G. and V.E. Yarsley, *Plastics in the Service of Man*, Penguin, Harmondsworth, 1956

Plastics, Pelican, Harmondsworth, 1941

Crone, Rainer, *Andy Warhol*, translated from the German (Hamburg, 1970), Thames and Hudson, London, 1970

Dent, Roger, *Principles of Pneumatic Architecture*, Architectural Press, London, 1971

Dingley, Cyril S., *The Story of B.I.P.* B.I.P. Ltd., Birmingham, 1962

Drexley, Arthur, *The Design Collection: Selected Objects*, Museum of Modern Art, New York, 1970

Furniture from the Design Collection, Museum of Modern Art, New York, 1973

DuBois, J. Harry, *Plastics*, Reinhold, New York, 1975

Plastics History—USA, Cahners, Boston, Mass., 1972

Dunlop Educational Division, *This is Where Rubber Begins*, Dunlop Co. Ltd, London, 1963

Feiffer, Jules, *The Great Comic Book Heroes*, Dial, New York, 1965; Allen Lane and Penguin, London, 1967

Fleck, H.R., *The Story of Plastics*, Burke, London, n.d.

Plastics—Scientific and Technological, English Universities Press, London, 1951

Glanville, Alan B., *Plastics and Society*, paper read at Plastics Institute National Technical Conference, 1972. Reprinted in *Plastics and Polymers*, October 1973

Gloag, John, *Plastics and Industrial Design*, Allen and Unwin, London, 1946

Hillier, Bevis, *Art Deco*, Studio Vista/Dutton Pictureback, London and New York, 1968

The World of Art Deco, Studio Vista, London, 1971

Howarth-Loomes, B.E.C., *Victorian Photography*, Ward Lock, London, 1974

ICI Ltd, *Landmarks in the Plastics Industry*, ICI Ltd Plastics Division, Welwyn Garden City, 1962

International Institute of Synthetic Rubber Producers Inc., *Synthetic Rubber—Story of an Industry*, New York, 1973

Johnson, Ellen H., *Claes Oldenburg*, Penguin New Art 4, Penguin, Harmondsworth, 1971

Kaufman, M., *The First Century of Plastics*, The Plastics and Rubber Institute, London, 1963

Kaufmann Jr, E., *Prize Designs for Modern Furniture*, Museum of Modern Art, New York, 1950

Leyson, Capt. Burr W., *Plastics in the World of Tomorrow*, Paul Elek, London, 1946

Lippard, Lucy, *Pop Art*, Thames and Hudson, London, 1966

Lushington, Roger, *Plastics and You*, Pan Books, London, 1967

Mark, Herman F., *Giant Molecules*, Time Life, Holland, 1970

Megson, N.J.L., *Plastics*, Longmans, Green and Co., London, New York, Toronto, 1948

Merriam, John, *Pioneering in Plastics, the Story of Xylonite*, East Anglian Magazine, Ipswich, 1976

Nelson, George, *Chairs*, Whitney, New York, 1953

Newman, Thelma, *Plastics as an Art Form*, Pitman, London, 1972

Papanek, Victor, *Design for the Real World*, Bantam, New York, 1973

Parkyn, Brian, *Glass Reinforced Plastics*, Illiffe, London, 1970

Penfold, R.C., *A Journalist's Guide to Plastics*, British Plastics Federation, London, 1969

Plastics and Rubber Institute, *Papers from the Plastics in Furniture Conference*, London, 1970

Sources of Information on Plastics and Rubber, London, 1976

Powell, P.C., *Plastics for Industrial Designers*, The Plastics and Rubber Institute, London, 1973

Product Journals Ltd, *Plastics—Design Engineering Handbook*, Kent, 1968

Quarmby, Arthur, *The Plastics Architect*, Pall Mall, London, 1974

Redfarn, C.A., *A Guide to Plastics*, Illiffe, London, 1958

Roukes, Nicholas, *Sculpture in Plastics*, Watson-Guptill, New York; Pitman, London, 1968

Saarinen, Eero, *Eero Saarinen on His Work*, Yale University Press, 1968

Society of the Plastics Industry, *Plastics and the Human Body*, papers from a symposium at the National Plastics Conference (Chicago, 1968), Cahners, Colorado, 1968

Smith, Paul I., *Practical Plastics*, Odhams, London, 1947

Telcon Plastics Ltd, *The Telcon Story 1850-1950*, Telegraph, Construction and Maintenance Co., London, 1950

Tooley, Peter, *High Polymers*, John Murray, London, 1971

Walker, John, *Glossary of Art, Architecture and Design since 1945*, Clive Bingley, London, 1973

Wordingham, J.A. and P. Reboul, *Dictionary of Plastics*, Newnes, London, 1964

Yarsley, V.E., *Plastics Applied*, The National Trade Press Ltd, London, 1945

Zanuso, Marco, *Dunhill Industrial Design Australian Lecture Series*, Trevor Miller Publishing, Melbourne, 1971

Journals and Periodicals

Abitare, Milan

The Architect's Journal, Architectural Press, London

Architectural Design, Standard Catalogue Co., London

Architectural Review, Architectural Press, London

Art and Artists, Hansom Books, London

Art Forum, Charles Cowles, New York

Art and Industry, The Studio, London

Beetle Bulletin, British Industrial Plastics, Oldbury

Building Design, Morgan-Grampian, London

Building Specification, Industrial Publications, Dublin

Design, Council of Industrial Design, London

Designer, Society of Industrial Artists and Designers, London

Domus, Milan

Encyclopaedia of Polymer Science and Technology, Interscience (Division of Wiley and Sons Inc.), New York, London, Sydney, Toronto

European Plastics News (incorporating *British Plastics and Europlastics Monthly*), I.P.C. Industrial Press, London

Plastforum, Sweden

House and Garden, Condé Nast, London

Industrial Design, Whitney, New York

La Maison de Marie Claire, Marie Claire Album, Paris

Materie Plastiche ed Elastomeri, Editrice L'Industria, Milan

Mobilia, Mobilia Productions, Snekkersten, Denmark

Modern Plastics Encyclopaedia, McGraw-Hill, New York

Modern Plastics International, McGraw-Hill, Lausanne

Plasticonstruction, Carl Hanser Verlag, Munich

Plastics and Polymers, Plastics and Rubber Institute, London

Plastics and Rubber Weekly, Maclaren, Croydon

Plastics Engineering, Society of Plastics Engineers, Connecticut

Plastics in Building, ICI Ltd, Welwyn Garden City

Plastics Technology, Bill Communications, New York

Plastics Today, ICI Ltd, Welwyn Garden City

Plastics World, Cahners, Boston, Mass.

Polymer Age, Rubber and Technical Press, Kent

Pulse Magazine, Morgan-Grampian, London

Studio International, London

Exhibition Catalogues

British Art in Industry, Royal Academy, London, 1935

Campo Vitale, Palazzo Grassi, Venice, 1967

Design 46, 'Britain Can Make It' Exhibition, Victoria and Albert Museum, London, 1946

Design and Plastics, Museum of Decorative Arts, Prague, 1972

Designing in Plastics, Design Centre, London, 1971

Modern Chairs, Whitechapel Art Gallery, London, 1970

Naum Gabo, Tate Gallery, London, 1966

Olivier Mourgue, Musée des Arts Décoratifs, Nantes, 1976

Plastic as Plastic, Museum of Contemporary Crafts, New York, 1968-9

Prospex 67, Royal College of Art, London, 1967

La Sedia in Materiale Plastico, Centrokappa, Milan, 1975

S'Asseoir, Musée de Grenoble, 1974

Siège-Poème, Maison des Arts et de la Culture, Créteuil, 1975

Index

Page numbers in italics refer to illustrations